Tissue Science Engineering and Technology

Tissue Science Engineering and Technology

Dr. Neeraj Tandan

RANDOM PUBLICATIONS
NEW DELHI (INDIA)

Tissue Science Engineering and Technology

ISBN 978-93-5111-374-4

Published in 2014 in India by

RANDOM PUBLICATIONS

4376-A/4B, Gali Murari Lal, Ansari Road
New Delhi-110 002
Phone : +91-11-43580356, +91-11-23289044
e-mail: randomexports@gmail.com, sales@randompublications.com, info@randompublications.com

Type Setting by : Keystoneprintads, Delhi-110051
Digitally Printed at : Replika Press Pvt. Ltd.

Preface

Tissue engineering is a multidisciplinary field incorporating the principles of biology, engineering, chemistry, and medicine to create natural substitutes of native tissues for scientific investigate or medical use. Specific submissions of this technology include the study of tissue expansion and function, investigating drug reaction, and tissue repair and replacement. This area is rapidly becoming one of the most promising conduct options for patient's affliction from tissue failure.

Tissue engineering is a relative new field of research aiming at repair of damaged tissues or organs. To achieve this purpose cells are cultured in a laboratory, then combined with a biomaterial with or without growth factors to produce a specific tissue. Matters such as the choice of cells - either tissue specific cells (differentiated cells) or stem cells (undifferentiated cells) - the consequences of the interactions of cells with biomaterials, the structure of the biomaterial and the cell survival after implantation, will be addressed.

Tissue engineering / regenerative medicine is an emerging multidisciplinary field involving biology, medicine, and engineering that is likely to revolutionize the ways we improve the health and quality of life for millions of people worldwide by restoring, maintaining, or enhancing tissue and organ function. In addition to having a therapeutic application, where the tissue is either grown in a patient or outside the patient and transplanted, tissue engineering can have diagnostic applications where the tissue is made in vitro and used for testing drug metabolism and uptake, toxicity, and pathogenicity. The foundation of tissue engineering/regenerative medicine for either therapeutic or diagnostic applications is the ability to exploit living cells in a variety of ways.

Cells from the biopsy are then cultured from explants or a collagenase digestion to create a "cell bank". These cells are then further cultured on collagenous substrates, under the correct physiological conditions, to form Tissue Engineered constructs for implantation. The process is carried out in a Tissue Culture facility to maintain a sterile environment. Cellular biochemical and physical activity can be enhanced by the addition of growth factors or cytokines, also by the use of physical stimulation. The tensioning-Culture Force

Monitor applies minute physical loads to stimulate the resident cell population in the collagenous scaffold into bio-chemical and bio-physical activity normally associated with organogenesis and tissue repair. After further tissue culture under the correct conditions, the resident cells in the Tissue Engineered construct will disolve the original collagen scaffold and secrete a new collagen rich neo-tissue, the construct can then be implanted back into the patient from whom the cells were originally removed.

This science will remove the necessity for anti-rejection drugs as the living tissue engineered construct has been grown from the patients own cells, and will be accepted as a natural part of the patients body. This book will be useful to biotechnologists, biologists, agriculture scientists, researchers, teachers and students of plant sciences.

I thank all members of my team who have helped in the preparation of the book. My special thanks go to "Random Publications" who have published the book.

– Dr. Neeraj Tandan

Contents

1

Introduction

TISSUE ENGINEERING

Tissue engineering is the use of a combination of cells, engineering and materials methods, and suitable biochemical and physio-chemical factors to improve or replace biological functions. While most definitions of tissue engineering cover a broad range of applications, in practice the term is closely associated with applications that repair or replace portions of or whole tissues (i.e., bone, cartilage, blood vessels, bladder, etc.). Often, the tissues involved require certain mechanical and structural properties for proper functioning.

The term has also been applied to efforts to perform specific biochemical functions using cells within an artificially-created support system (e.g. an artificial pancreas, or a bioartificial liver). The term regenerative medicine is often used synonymously with tissue engineering, although those involved in regenerative medicine place more emphasis on the use of stem cells to produce tissues.

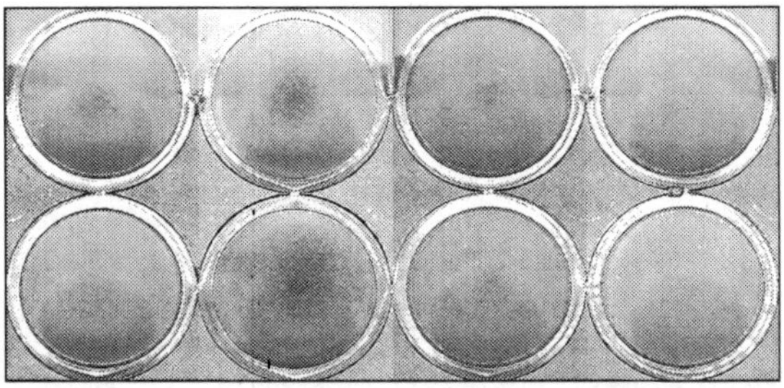

A commonly applied definition of tissue engineering, as stated by Langer and Vacanti, is "an interdisciplinary field that applies the principles of engineering and life sciences toward the development of biological substitutes that restore, maintain, or improve tissue function or a whole organ". Tissue engineering has also been defined as "understanding the principles of tissue growth, and applying this to produce functional replacement tissue for clinical use." A further description goes on to say that an "underlying supposition of tissue engineering is that the employment of natural biology of the system will allow for greater success in developing therapeutic strategies aimed at the replacement, repair, maintenance, and/or enhancement of tissue function." Powerful developments in the multidisciplinary field of tissue engineering have yielded a novel set of tissue replacement parts and implementation strategies. Scientific advances in biomaterials, stem cells, growth and differentiation factors, and biomimetic environments have created unique opportunities to fabricate tissues in the laboratory from combinations of engineered extracellular matrices ("scaffolds"), cells, and biologically active molecules.

Among the major challenges now facing tissue engineering is the need for more complex functionality, as well as both functional and biomechanical stability in laboratory-grown tissues destined for transplantation. The continued success of tissue engineering, and the eventual development of true human replacement parts, will grow from the convergence of engineering and basic research advances in tissue, matrix, growth factor, stem cell, and developmental biology, as well as materials science and bioinformatics. The NSF published a report entitled "The Emergence of Tissue Engineering as a Research Field", which gives a thorough description of the history of this field.

Examples

- Bioartificial liver device — several research efforts have produced hepatic assist devices utilizing living hepatocytes.
- Artificial pancreas — research involves using islet cells to produce and regulate insulin, particularly in cases of diabetes.
- Artificial bladders — Anthony Atala has successfully implanted artificially grown bladders into seven out of approximately 20 human test subjects as part of a long-term experiment.
- Cartilage — lab-grown tissue was successfully used to repair knee cartilage.
- Doris Taylor's heart in a jar
- Tissue-engineered airway
- Artificial skin constructed from human skin cells embedded in collagen
- Artificial bone marrow.

HYBRID BIOMATERIALS FOR BONE TISSUE ENGINEERING

Tissue engineering has evolved from the use of biomaterials for bone substitution that fulfill the clinical demands of biocompatibility, biodegradability, non-immunogeneity, structural strength and porosity. Porous scaffolds have been developed in many forms and materials, but few reached the need of adequate physical, biological and mechanical properties. In the present paper we report the preparation of hybrid porous polyvinyl alcohol (PVA)/bioactive glass through the sol-gel route, using partially and fully hydrolyzed polyvinyl alcohol, and perform structural characterization. Hybrids containing PVA and bioactive glass with composition $58SiO_2$-33CaO-$9P_2O_5$ were synthesized by foaming a mixture of polymer solution and bioactive glass sol-gel precursor solution. Sol-gel solution was prepared from mixing tetraethoxysilane (TEOS), triethylphosphate (TEP), and calcium chloride as chemical precursors.

The hybrid composites obtained after aging and drying at low temperature were chemically and morphologically characterized through infrared spectroscopy and scanning electron microscopy. The degree of hydrolysis of PVA, concentration of PVA solution and different PVA-bioglass composition ratios affect the synthesis procedure. Synthesis parameters must be very well combined in order to allow foaming and gelation. The hybrid scaffolds obtained exhibited macroporous structure with pore size varying from 50 to 600 μm.

Tissue engineering represents a new field that aims to grow complex, three-dimensional tissues or organs to replace damaged tissues. For this purpose, investigation of novel biomaterials for bone engineering represents an essential area for planning tissue engineering approaches. Next generation biomaterials should combine bioactive and bioresorbable properties to activate in vivo mechanisms of tissue regeneration, stimulating the body to heal itself and to facilitate replacement of the scaffold by the regenerating tissue. A variety of biomaterials, including synthetic polymers, ceramics and natural polymers are being used to fabricate synthetic scaffold that acts as guide and stimulus for the three-dimensional tissue growth.

Controlled porous architecture, high porosity, adequate pore size and interconnectivity, of these scaffolds is necessary to facilitate cell seeding and diffusion throughout the whole structure of both cells and nutrients. The minimum pore diameter required for bone ingrowth and angiogenesis into a scaffold is considered to be 100 μm. Previous in vitro work has suggested that the ideal pore diameter for bone ingrowth is between 300-400 μm. The material must exhibit good biocompatibility meaning that the material must not demonstrate severe immunogenicity or cytotoxicity. The ability to control scaffold degradation and mechanical integrity is also a requirement to be fulfilled in the development of optimal scaffolds. Ideally, scaffolds should closely match the properties of the tissue it is to replace.

Either biodegradable polymer or ceramic scaffolds, commonly being considered for bone tissue engineering, lack adequate mechanical properties. Synthetic bioresorbable polymers are easily fabricated into complex structures, yet they are too weak to meet the demands of surgery and the in vivo physiological environment. Conversely, ceramic scaffolds have low toughness and strength. Efforts have been made in developing composite of polymers and ceramics with the aim to increase the mechanical stability and improve tissue interaction. The material composition might give favorable mechanical properties, strength via ceramic phase, toughness and plasticity via polymer phase. Thus, hybrid materials derived from the integration of biodegradable polymers with bioactive inorganic material might be promising on scaffolds development. The bioactive ceramic phase may be a calcium phosphate ceramic, a glass-ceramic or a bioactive glass.

Bioactive glasses are able to form a bond to living bone, which has been attributed to the formation of hydroxyl carbonate apatite layer under physiological conditions. However bioactive glass is a stiff and brittle material, difficult to form into complex shapes and prone to catastrophic fracture under loads.

The sol-gel method has been employed for the synthesis of hybrid organic-inorganic composite materials. Sol-gel derived bioactive glass foams were also successfully obtained. In vitro cell studies in the presence of these foams have shown an increase in osteoblast proliferation and collagen production as well as the stimulation of the formation and mineralisation of bone nodules reinforcing their potential. However, their compressive strength is in the range of 0.3 to 2.5 MPa, depending on sintering temperature and final pore structure, a value lower than even trabecular bone.

Nevertheless, it is in the range obtained for several scaffolds for tissue engineering reported in the literature. However the toughness and tensile strength of the foams are lower than those of bone. It is expected that the preparation of polymer bioactive glass hybrid foams may lead to scaffolds with better mechanical behaviour compared to pure bioactive glass foams. It should be emphasized however that these materials are expected to be used in sites free of dynamic load, or with controlled load during tissue regeneration.

Polyvinyl alcohol, PVA, is a water soluble synthetic resin which is obtained through polymerization of vinyl acetate monomer. By hydrolysis, the acetate groups are converted in hydroxyl groups. The degree of hydrolysis in a polyvinyl alcohol reagent is controlled by this process. The polar nature of poly (vinyl alcohol) facilitates the formation of hydrogen bonds and eventual condensation with silanol groups (from developing polysilicate network) formed by hydrolysis of the silicon alkoxides.

Moreover PVA has been proposed for controlled release systems and is employed in a variety of biomedical applications, generally being considered to be biocompatible. Although not a biodegradable polymer itself, when used with molecular weight less than around 10.000 g/mol and associated with a biodegradable sol-gel derived bioactive glass, PVA molecules are expected to be eliminated by the body.

Synthetic bioactive and bioresorbable composite materials are becoming increasingly important as scaffolds for tissue engineering. Bioactive glass reacts with physiological fluids to form tenacious bonds to hard (and in some cases soft) tissue. Composites of tailored physical, biological and mechanical properties as well as predictable degradation behaviour can be produced combining bioresorbable polymers and bioactive inorganic phases. Both PVA, as well as bioactive glasses, are being currently used as biomaterials, candidates for tissue engineering applications.

In previous work of our group the synthesis of PVA/silica hybrids obtained via sol-gel method was reported. The hybrids were prepared with low polymer contents (20-40 wt. (per cent)). In this work, we present results concerning the sol-gel synthesis of polyvinyl alcohol-bioactive glass hybrid foams, with polymer contents from 60 to 80 wt. (per cent), and different types of PVA. The purpose of this study was to investigate the effects of some important synthesis parameters such as degree of hydrolysis of PVA, concentration of PVA solution and higher PVA content on the final properties of composites.

As far as we know, this is the first research in this technology field to address the influence of PVA chemical properties such as degree of hydrolysis on the fabrication of bioactive glass-polymer (PVA) hybrid tri-dimensional scaffolds.

Experimental

Hybrids Preparation

Sol-gel derived bioactive glass/polyvinyl alcohol hybrid foams were prepared using a procedure similar to the one used by Pereira et al.

Preparation of PVA Solution

The Polyvinyl alcohol (PVA) selected for use were: from Celanese Chemicals, Celvol 103, degree of hydrolysis 98.0-98.8 per cent, molecular weight range: 13,000-23,000 g/mol and from Aldrich-Sigma, degree of hydrolysis 80 per cent, molecular weight: 9,000-10,000 g/mol.

PVA aqueous solutions were prepared at 23, 28 and 35 wt. (per cent) by dissolving the PVA powder in a water bath at 80°C, under constant stirring, for 2 hours. The pH of the solution was adjusted to 2.0 by hydrochloric acid solution 2 N.

Preparation of the Starting Bioglass Solution

The starting sol solution with composition 58 wt. (per cent) SiO_2- 33 wt. (per cent) CaO- 9 wt. (per cent) P2O5 was synthesized from mixing tetraethoxysilane (TEOS), D.I. water, triethylphosphate (TEP), and calcium chloride in presence of hydrochloric acid solution 2 N. The H2O/TEOS molar ratio used was 12.

Preparation of the PVA-Bioactive Glass Hybrid Foams

The hybrid compositions prepared were 20, 30, 40 wt. (per cent) glass and 80, 70, 60 wt. (per cent) polymer. The hybrids were obtained using a procedure similar to the one described by in our previous work. An appropriate amount of the starting sol was added to the PVA solution. To this resulting solution were added the surfactant, sodium laureate sulfate (LESS), and HF 10 per centv/v solution.

The mixture was foamed by vigorous agitation. Hydrofluoric acid was used to catalyze the gelation. The foams were cast just before gelation in plastic containers and sealed. The samples were, basically, aged at 40°C for 3 days and then dried at 40°C for 7 days.

Hybrids Characterization

FTIR was used to characterize the presence of chemical groups. The FTIR absorption spectra were obtained within the range between 4000 and 400 cm-1 (Perkin Elmer, Paragon 1000), using the KBr pellet method. For the PVA sample was prepared a thin film and used the transmission spectroscopy. Porous morphologies of the hybrids were examined with a scanning electron microscope (JEOL, JSM-6360L) after carbon coating.

The examination was carried out at accelerating voltage of 15 kV. To expose the internal architecture, the samples were cut with a razor blade. Hybrid PVA/bioglass foams were obtained for polymer contents of 60, 70 and 80 wt. (per cent). The use of two degree of hydrolysis of PVA changed the solution behaviour.

For PVA, fully hydrolyzed (Celvol 103), the concentrations of the starting PVA solution that proved to be adequate to allow the production and gelation of the foam were 23 and 28 wt. (per cent). Using 23 wt. (per cent), the resulting final foams have less residual water and the drying procedure is more efficient, compared to lower concentration PVA solutions used in previous work. Comparing with the 28 wt. (per cent) concentration, the amount of hydrofluoric acid necessary to gel the foam was higher.

Reaching gelation can be better controlled when working with PVA fully hydrolyzed, Celvol 103, since its solution has higher viscosity and the higher predominance of hydroxyl groups may contribute to crosslink with the polysiloxane network, through hydrogen bonds and condensation with silanol groups.

For PVA, 80 per cent hydrolyzed, the starting PVA concentration of 23 wt. (per cent) did not work, even for higher HF concentration used foams could not be stabilized, since the gel-sol transition could not be reached using the procedure adopted.

At 28 wt. (per cent), depending of the hybrid composition, it was possible obtain the hybrid foams. For hybrid composition of PVA higher than 70 wt. (per cent), the control of the foam casting could not be done, since the gel-sol transition is a very slow process in the conditions used. PVA, 80 per cent hydrolyzed, presents higher amount of acetate groups left in the backbone, providing less reaction sites for reaction with silanol groups formed during hydrolysis of silicon precursors.

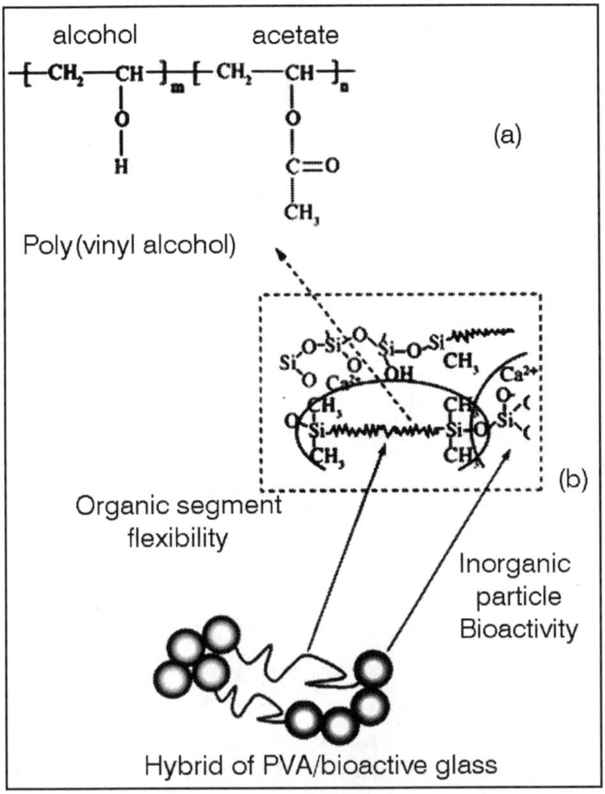

Fig. Schematic Drawing of the Inorganic -Organic Hybrid synthesized Based on PVA Polymer and Bioactive Glasses; a) PVA Chain with Variable Hydrolysis Degree; Hybrid Network Structure

The presence of acetate groups in partially hydrolyzed polymers weakens the hydrogen bonding. This suggests that the mechanism involving interaction between the hydroxyl groups of PVA and hydroxyl groups of silanol, from hydrolyzed silica, is delayed. Chemical crosslinking, is a versatile method to modify and improve properties in materials.

In this way, there are several advantages on using organic-inorganic hybrid, because one may be able to tailor physical, biological and mechanical properties as well as predictable degradation behaviour where novel materials can be produced combining bioresorbable polymers and bioactive inorganic phases.

FTIR Characterization

Representative IR spectra of the pure PVA films, the bioglass and the hybrid foams, for the sample group of PVA fully hydrolyzed, and the PVA 80 per cent hydrolyzed sample group. The spectra of the other hybrids obtained in this work are very similar to the spectra exhibited.

No relevant differences were found by using the two PVA precursors neither for varying the PVA solution and composition of the hybrids.

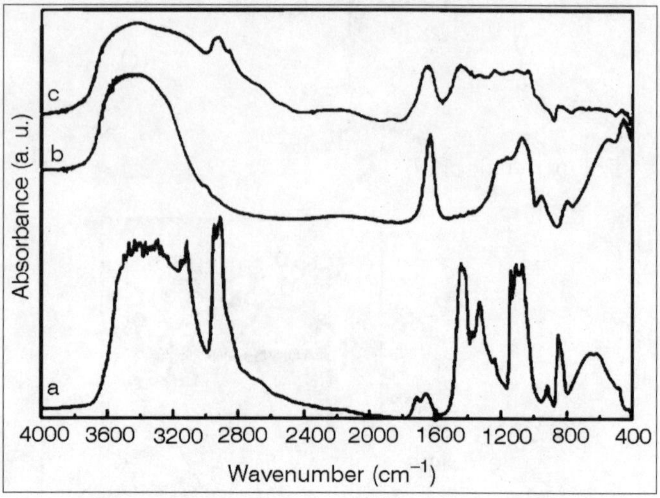

Fig. Ftir Spectra of Samples: a) Polyvinyl Alcohol, Fully Hydrolyzed, Celvol 103; b) Bioglass; c) Hybrid 70 wt. (per cent) PVA -30 wt. (per cent) Bioglass (3B-7Pc).

FTIR spectra of pure PVA are shown. For both PVA spectra, the broad band observed from 3100 to 3600 cm-1 may be assigned to O-H stretching due the strong hydrogen bond of intramolecular and intermolecular type. The presence of a higher hydroxyl content in the fully hydrolyzed PVA compared to the PVA 80 per cent hydrolyzed can be seen from the broader band around 3300 cm^{-1}. The C-H alkyl stretching band can be observed at 2850-2950 cm^{-1}.

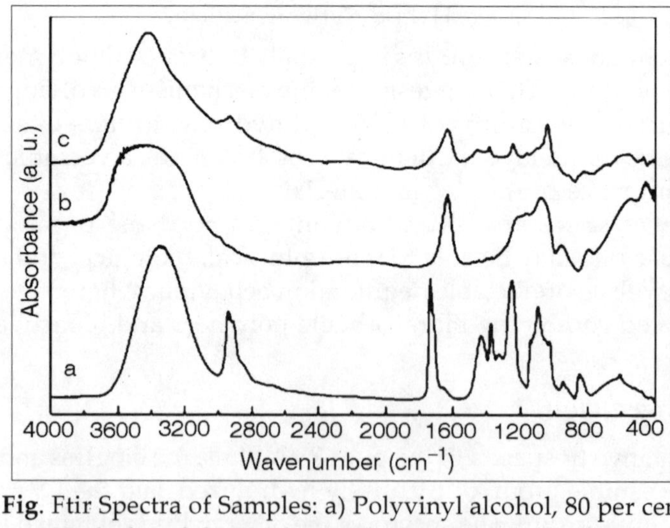

Fig. Ftir Spectra of Samples: a) Polyvinyl alcohol, 80 per cent Hydrolyzed; b) Bioglass; c) Hybrid 70 wt. (per cent) PVA -30 wt. (per cent) Bioglass (3B-7Ph)

The absorption peaks at approximately 1710 cm^{-1} (for fully hydrolyzed PVA) or 1740 cm-1 (for 80 per cent hydrolyzed) and at 1090-1150 cm^{-1} may be attributed to the stretching vibration of C = O and C-O of the remaining vinyl acetate non-hydrolyzed group of PVA polymer. The absorption band around 1700 cm^{-1} arises due to the carbonyl band (C= O) of the acetate group found in partially hydrolyzed PVA polymer and has higher relative intensity for the PVA 80 per cent hydrolyzed as expected from the higher concentration of remaining acetate groups.

In the FTIR spectra of the bioglass presented in both figures, the vibrational modes due to Si-O-Si asymmetric stretching are observed at approximately 1070 cm-1 and 1200 cm-1. Additional peaks are seen at 790 cm^{-1} (for symmetric Si-O-Si stretching vibration, and at 450 cm^{-1} (for Si-O-Si bending modes). A typical absorption band observed in silica gel is located at 1640 cm^{-1} and is attributed to the deformation mode of adsorbed molecular water in the pores. The band at 950 cm^{-1} has been ascribed to the stretching vibration of silanol groups in pure silica (vibrational modes of Si-O- stretching).

FTIR spectra of hybrid in the composition 70 wt. (per cent) PVA30 wt. (per cent) bioglass are presented for Celvol and 80 per cent hydrolyzed, respectively. The bioglass component is identified by major vibration bands, (Si-O-Si, at 1080 and 450 cm^{-1}). For both hybrids, the peak at 950 cm^{-1} associated with the Si-OH vibrational mode remains as a shoulder. In the frequency range of 3000-3650 cm^{-1}, a broader band was noted, especially for hybrid prepared from PVA fully hydrolyzed.

The band located at 1640 cm-1 remains in the spectra of the hybrids since the low heating temperature is not enough to remove the molecular water of the pores. The terminal vinyl groups of the PVA which appear at 1710 cm^{-1} (for fully hydrolyzed PVA) or 1740 cm^{-1} (for 80 per cent hydrolyzed) are not shown in the hybrids, which suggests that these bonds might be involved in crosslinking with silica. In the range 1500-900 cm^{-1} there is a superposition of the bands derived from the bioglass and the PVA components.

Scanning Electron Microscopy

The pore morphology and distribution of the hybrid foams can be seen by the SEM micrographs presented in Figure. Figure shows SEM micrograph of hybrid 70 wt. (per cent) PVA-30 wt. (per cent) Bioglass (3B-7Pc), SEM micrograph of hybrid 80 wt. (per cent) PVA-20 wt. (per cent) Bioglass (2B-8Pc), Figure shows SEM micrograph of hybrid 70 wt. (per cent) PVA-30 wt. (per cent) Bioglass (3B-7Ph 35), and Figure shows the micrograph of 60 wt. (per cent) PVA-40 wt. (per cent) Bioglass (4B-6Ph-35).

Very distinct features can be observed in the different hybrids produced. The observed morphologies of the foams vary considering the processing variables. The porosity and pore sizes are strongly dependent on the hybrid composition, concentration of the polymer solution used and gelation time.

The control of the gelation time is very important to establish an adequate morphology of the hybrids produced by this method. For composition of 70 wt. (per cent) PVA, for both PVA polymers used, the gelation time was found to be approximately 13 minutes and this is a proper time to gel the three-dimensional hybrids structure and stabilize the bubbles formed during the process.

These hybrids exhibited macroporous structure with interconnected open pores and pore size varied from 50 to 350 mm. For the composition of 60 wt. (per cent) PVA, the gelation time was shorter and the time at which the foam was cast, the gelation point almost reached, affecting the final porous structure.

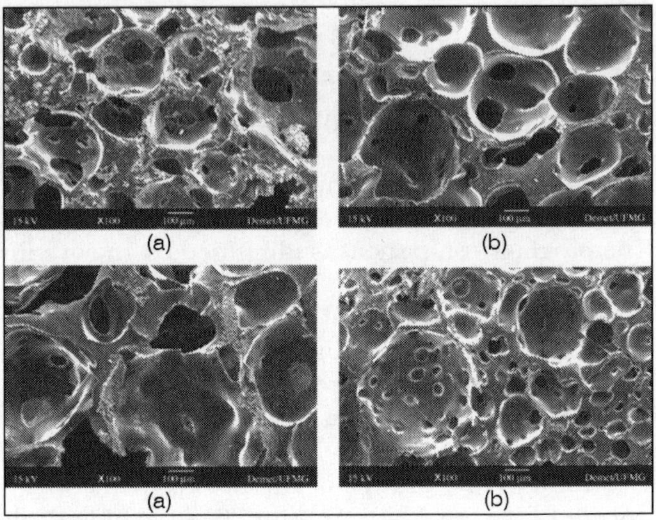

Fig. Scanning Micrographs of Hybrids: a) 70 wt. (per cent) PVA -30 wt.(per cent) Bioglass (3B-7Pc); b) 80wt. (per cent) PVA -20wt. (per cent) Bioglass (2B-8Pc); c) 70wt. (per cent) PVA -30wt. (per cent) Bioglass (3B-7Ph-35); d) 60wt. (per cent) PVA -40wt. per cent) Bioglass (4B-6Ph-35)

For this hybrid, the distribution of pores is more homogeneous and pore size ranged from 50 to 90 mm and reaching 450 mm for the bubble-like pores. For the composition of 80 wt. (per cent) PVA, For both PVA polymers used, the gelation time was delayed and the pore morphology was semi-spherical. Hybrid showed in Figure has the largest pores (pore size from about 160 to 500 mm).

The number of possible combinations of bioactive glasses with biocompatible polymers such as PVA is almost countless. That means each and every proportion of inorganic phase CaO, SiO_2 and P_2O_5, not to mention the synthesis parameters themselves, will result on a different hybrid system with some specific properties and bioactive behaviour. Therefore, synthetic bioactive and bioresorbable composite materials are becoming increasingly important as scaffolds for tissue engineering. Next generation biomaterials must combine bioactive and bioresorbable properties to activate in vivo mechanisms of tissue regeneration, stimulating the body to heal itself and to facilitate replacement of the scaffold by the regenerating tissue.

Bioactive glass reacts with physiological fluids to form tenacious bonds to hard (and in some cases soft) tissue. However bioactive glass is a stiff and brittle material, difficult to form into complex shapes and prone to catastrophic fracture under loads. Conversely, synthetic bioresorbable polymers are easily fabricated into complex structures, yet they are too weak to meet the demands of surgery and the in vivo physiological environment. Composites of tailored physical, biological and mechanical properties as well as predictable degradation behaviour can be produced combining bioresorbable polymers and bioactive inorganic phases.Hybrid Polyvinyl alcohol/bioglass foams have been prepared by the sol-gel method, with a high content of polymer (60 to 80 per cent).

The synthesis parameters need to be very well combined in order to allow the control of foaming and gelation. Pore characteristics of the obtained scaffolds have been related to hybrid composition, concentration of the polymer solution used, and gelation time. The macropore diameter ranged from 50 to 600 mm. The results suggest that there is scope for designing materials with controlled compositions and morphologies. Such system has a potential to be used in bone tissue engineering, but further studies must be conducted.

BONE TISSUE ENGINEERING USING FOETAL CELL THERAPY

Different cell sources for bone tissue engineering are reviewed. In particular, adult cell source strategies have been based on the implantation of unfractionated fresh bone marrow; purified, culture expanded mesenchymal stem cells, differentiated osteoblasts, or cells that have been modified genetically to express rhBMP. Several limiting factors are mentioned for these strategies such as low number of available cells or possible immunological reaction of the host.

Foetal bone cells are presented as an alternative solution and a review of actual treatments using these cells is presented. Finally, foetal cells used specifically for bone tissue engineering are characterised and potentially interesting therapeutic options are proposed. Bone regeneration is based on the hypothesis that healthy progenitor cells, either recruited or delivered to an injured site, can ultimately regenerate lost or damaged tissue. Three-dimensional bone grafts may enhance bone regeneration by creating and maintaining a space that facilitates progenitor cell migration, proliferation and differentiation. Several techniques have been developed to provide the surgeon with needed material for bone grafts. Pieces of bone are collected either from the patients undergoing surgery (autograft, which is considered as the gold standard and remains the most used therapy for bone repair) or obtained from a tissue bank (allograft). However, these techniques have some drawbacks such as traumatic procedures, morbidity or increased operating times for autografts and limited supply, risks of contamination or high costs for allografts.

These limitations have led to new research aiming to provide a bone graft engineered in the laboratory and available when needed by the surgeon. The ultimate goal of this new treatment strategy is the regeneration rather than just the repair of skeletal tissue. From a broad point of view, this approach can be defined as "bone tissue engineering". As for any application in tissue engineering, the cell origin and type are essential aspects in bone tissue engineering. Technically, the cells used should be relatively simple with respect to their collection, culture, expansion and storage. Therapeutically, the cells used should have high bone formation potential, no immunological-induced reactions and no pro-inflammatory properties.

ADULT CELL SOURCES FOR BONE TISSUE ENGINEERING APPLICATIONS

Cell sources for bone tissue engineering applications can be categorised with respect to their state of differentiation. With this idea, four different cell-based tissue engineering approaches have been described for the regeneration of bone. These strategies are based on the implantation of:

- Unfractionated fresh bone marrow,
- Purified, culture expanded mesenchymal stem cells.
- Differentiated osteoblasts,
- Cells that have been modified genetically to express rhBMP.

Generally, the less differentiated cells will be more easily expanded in vitro due to their high proliferation rate, while the differentiated cells will be more effective in vivo due to their higher production of mineralised extracellular matrix.

For each type of cells used, advantages and disadvantages can be found. The cells from unfractionated fresh bone marrow are relatively easy to collect, but it will not be possible to use these cells in allotransplantation as bone marrow contains T lymphocytes that encounter and respond to host antigens in virtually all tissues in the body, leading to multi-system graft-versus-host syndrome. Mesenchymal stem cells, isolated from bone marrow aspirate, adult peripheral blood, neonatal cord blood or liver for example, could present advantages from an immunological point of view.

However, as one of every 100,000 nucleated cells derived from bone marrow is a stem cell, a procedure of isolation is required in order to decrease the volume of material injected. Compared to unfractionated bone marrow, mesenchymal stem cells have been shown to generate greater bone formation in preclinical studies. However, gradual loss of both their proliferative and differentiation potential has been observed during in vitro expansion. To overcome this difficulty, telomerase therapy (hTERT) has been recently used and effectively extended stem cell life-span while maintaining or even enhancing their osteogenic potential.

Alternative sources of autologous precursor cells were investigated in fat by liposuction or in skeletal muscle by biopsy. It has to be mentioned that pluripotential cells from mesodermal tissues other than bone marrow did not show bone healing in animal models. The use of predifferentiated osteoblasts has been shown to enhance the rate and extent of bone regeneration. Their expansion in vitro may be however difficult and is highly depending on the donor. Finally, cells genetically modified to express bone formation cytokines could be used to take advantage of genetic therapy, combining gene therapy and tissue engineering methodologies to enhance tissue regeneration.

The transfection of the cells can be done either in two steps with an in vitro transfection followed by injection of the modified cells (procedure called ex vivo gene therapy) or in one step by transfecting directly the cells in the body (procedure called in vivo gene therapy). With ex vivo gene therapy approach, cells overexpressing BMP have been developed and used in animal studies. Concerns on possible host immune reaction need to be clarified. A list of requirements for ideal cell sources for bone tissue engineering applications is summarised in table. No adult cells seem to fulfil all the requirements leaving room to propose an alternative approach based on foetal cells.

FOETAL CELL SOURCES FOR TISSUE ENGINEERING APPLICATIONS

Foetal associated tissues such as placenta, amniotic liquid or umbilical cord are described to be potential sources of cells for tissue engineering. In contrast to embryonic tissue derived up to the end of the 8th week, foetal tissue begins at the 9th week and is considered as an organ donation. The idea of using foetal tissues or cells for cell therapy is mainly coming from the field of neurology. Indeed, several clinical studies have already been performed to treat neurodegenerative disorders. Neuronal affections such as Huntington's or Parkinson's disease have been treated by transplantation of fresh foetal neuroblasts. Unfortunately, these cells are difficult to expand in culture and have to be transplanted freshly therefore needing large quantities of fresh tissue.

Still within the neurology but following accidents, cell therapy for stroke has also tried to benefit from foetal cells, with a clinical trial treating patients with foetal cells obtained from the porcine primordial striatum. In this situation, safety of the treatment was verified, but none of the patients showed improvement. Treatment of spinal cord injury also tested the use of intraspinal foetal central nervous system grafts. Three clinical studies have been performed where the results mainly focused on security of the procedure. However, encouraging results were obtained. In particular translational correspondence was observed between preclinical locomotor performance in a cat before and after intraspinal foetal transplant surgery compared with a human subject before and after human foetal cell grafting. Transplantation of foetal central nervous system tissue is considered actually as a gold standard in neurobiology. Beside neurology, human foetal liver cells have been used already more than 10 years for transplantation to treat severe immunodeficiencies, haematological disorders and inborn errors of metabolism when there was no perfectly matched donor for marrow transplantation.

Recently, human foetal skin cells derived from one cell bank (1–4 cm^2 tissue results in over 10.5 million foetal skin constructs) were used in clinical trials and new advances in tissue therapy are possible with cellular constructs obtained from ex vivo cultures. Engineered regeneration of human skeletal adult tissues could be also developed using human foetal bone cells. Surprisingly, tissue engineering using foetal cells has barely begun to be investigated.

To evaluate their potential integration in a bone engineering strategy and following the experience gained in neurobiology, a biological characterisation of these cells is necessary, especially by evaluating the potential of foetal cells to produce mineralised bone extracellular matrix. Lately, a study was performed to specifically study the characteristics of human primary foetal bone cells for a better comprehension of their biology in vitro and to evaluate their potential use for tissue engineering in comparison to adult bone cells and mesenchymal stem cells.

Human primary foetal bone cells were compared to adult bone cells and mesenchymal stem cells for their ability to proliferate and to differentiate into osteoblasts in vitro. Cell proliferation, gene expression of bone markers, alkaline phosphatase (ALP) activity and mineralisation were analysed during a time-course study. Human primary foetal bone cells were compared to osteoblasts and mesenchymal stem cells obtained from adult tissues for their ability to proliferate and to differentiate into osteoblasts in vitro.

The doubling time of foetal bone cells was comparable to mesenchymal stem cells but significantly shorter than for adult osteoblasts. Gene expression of cbfa-1, ALP, a1 chain of type I collagen and osteocalcin were upregulated in foetal bone cells after 12 days of treatment with osteogenic factors, showing higher inductions than for adult osteoblasts and mesenchymal stem cells. The increase of ALP enzymatic activity was stronger for foetal than for adult osteoblasts reaching a maximum at day 10, but lower than for mesenchymal stem cells. Importantly, the mineralisation process of bone foetal cells started earlier than adult osteoblasts and mesenchymal stem cells. The human primary foetal bone cells have the advantages of high proliferation rate as for mesenchymal stem cells and effective production of mineralised extracellular bone matrix as for adult osteoblasts. Human primary foetal bone cells represent an interesting and promising potential for therapeutic use in the bone tissue engineering field as these cells can be easily stocked "frozen for use" when necessary.

COMPOSITE SCAFFOLDS FOR BONE TISSUE ENGINEERING

Since the late 1980s, tissue engineering (TE) as a new discipline has made rapid advances. Tissue engineering, as a potential medical treatment, holds promises of:

- Eliminating re-operations by using biological substitutes,
- Using biological substitutes to solve problems of implant rejection, transmission of diseases associated with xenografts and allografts, and shortage in organ donation,
- Providing long-term solutions in tissue repair or treatment of diseases,
- Potentially offering treatments for medical conditions that are currently untreatable.

Tissue engineering has therefore attracted great attention in science, engineering, medicine, and the society. It is now generally recognised that one of the key issues in tissue engineering is the development of suitable biodegradable materials and scaffolds for seeding cells and for the subsequent growth of tissues.

For bone tissue engineering, major issues of tissue engineering scaffolds include the use of appropriate matrix materials for scaffolds, control of porosity and pore characteristics of scaffolds, mechanical strength of scaffolds, scaffold degradation properties, and bioactivity (i.e., osteoconductivity, or osteoinductivity if possible) of scaffolds.

Biomaterials experience with biodegradable polymers in the pre-TE time has heavily guided researchers towards using medical-profession-accepted and the US Food and Drug Administration (FDA)-approved biopolymers such as poly(lactic acid) (PLA) and poly(lactic acid-co-glycolic acid) (PLGA) for constructing tissue engineering scaffolds. In recent years, other polymers have also been developed for tissue engineering applications.

Despite their phenomenal developments in the past three decades, bioactive bioceramics (including glasses and glass-ceramics) have been overlooked for tissue engineering applications until recently. It has been shown that some bioactive glasses have the potential for bone tissue engineering and there are now reports on the development of tissue engineering scaffolds purely made of bioceramics. Compared to the strengths of metals and ceramics for medical applications, the strengths of biodegradable polymers are already very low. With the introduction of pores in the polymers to form tissue engineering scaffolds, the strengths of porous structures are further decreased, as strengths of materials decrease drastically with an increase in porosity. On the other hand, polymers such as PLA and poly (PCL) are considered to be non-osteoconductive. In order to seek better scaffolds for bone tissue engineering, a composite strategy can therefore be adopted. Polymer-based scaffolds containing bioactive bioceramics can be produced in which the bioceramics can serve two purposes:

- Making the scaffolds osteoconductive,
- Reinforcing the scaffolds.

With this composite strategy, there are two approaches for making bioceramic-polymer composite scaffolds:

- Incorporating bioceramic particles in the scaffold through a variety of techniques;
- Coating a polymer scaffold with a thin layer of apatite through biomimetic processes.

Both strategies have been employed in our efforts to develop usable scaffolds for bone tissue engineering.

CELLS AS BUILDING BLOCKS

Tissue engineering utilizes living cells as engineering materials. Examples include using living fibroblasts in skin replacement or repair, cartilage repaired with living chondrocytes, or other types of cells used in other ways. Cells became available as engineering materials when scientists at Geron Corp. discovered how to extend telomeres in 1998, producing immortalized cell lines. Before this, laboratory cultures of healthy, noncancerous mammalian cells would only divide a fixed number of times, up to the Hayflick limit.

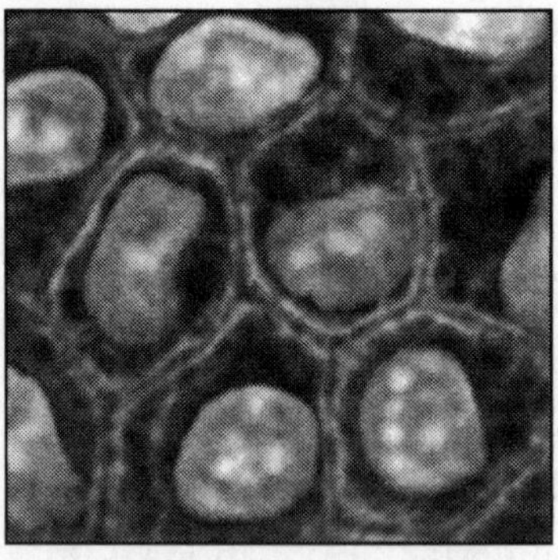

Fig. Stained Cells in Culture

Extraction

From fluid tissues such as blood, cells are extracted by bulk methods, usually centrifugation or apheresis. From solid tissues, extraction is more difficult. Usually the tissue is minced, and then digested with the enzymes trypsin or collagenase to remove the extracellular matrix that holds the cells. After that, the cells are free floating, and extracted using centrifugation or apheresis. Digestion with trypsin is very dependent on temperature. Higher temperatures digest the matrix faster, but create more damage. Collagenase is less temperature dependent, and damages fewer cells, but takes longer and is a more expensive reagent.

Types of Cells

Cells are often categorized by their source:

- Autologous cells are obtained from the same individual to which they will be reimplanted. Autologous cells have the fewest problems with rejection and pathogen transmission, however in some cases might not be available. For example in genetic disease suitable autologous cells are not available. Also very ill or elderly persons, as well as patients suffering from severe burns, may not have sufficient quantities of autologous cells to establish useful cell lines. Moreover since this category of cells needs to be harvested from the patient, there are also some concerns related to the necessity of performing such surgical operations that might lead to donor site infection or chronic pain. Autologous cells also must be cultured from samples before they can be used: this takes time, so autologous solutions may not be very quick. Recently there has been a trend towards the use of mesenchymal stem cells from bone marrow and fat. These cells can differentiate into a variety of tissue types, including bone, cartilage, fat, and nerve. A large number of cells can be easily and quickly isolated from fat, thus opening the potential for large numbers of cells to be quickly and easily obtained. Several companies have been founded to capitalize on this technology, the most successful at this time being Cytori Therapeutics.

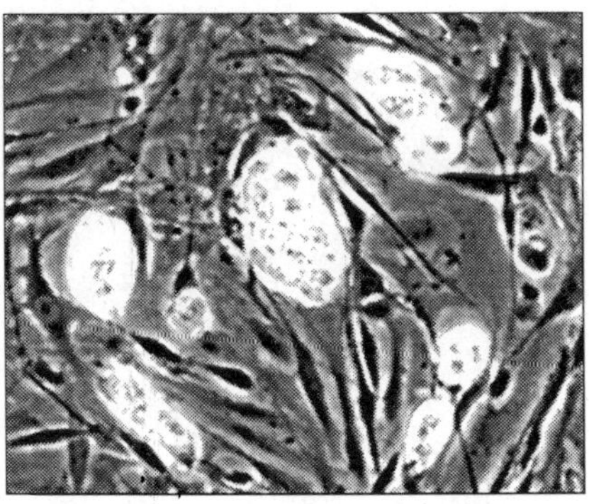

Fig. Mouse Embryonic Stem Cells. More Lab Photos

- Allogenic cells come from the body of a donor of the same species. While there are some ethical constraints to the use of human cells for in vitro studies, the employment of dermal fibroblasts from human foreskin has been demonstrated to be immunologically safe and thus a viable choice for tissue engineering of skin.
- Xenogenic cells are these isolated from individuals of another species. In particular animal cells have been used quite extensively in experiments aimed at the construction of cardiovascular implants.
- Syngenic or isogenic cells are isolated from genetically identical organisms, such as twins, clones, or highly inbred research animal models.
- Primary cells are from an organism.
- Secondary cells are from a cell bank.
- Stem cells are undifferentiated cells with the ability to divide in culture and give rise to different forms of specialized cells. According to their source stem cells are divided into "adult" and "embryonic" stem cells, the first class being multipotent and the latter mostly pluripotent; some cells are totipotent, in the earliest stages of the embryo. While there is still a large ethical debate related with the use of embryonic stem cells, it is thought that stem cells may be useful for the repair of diseased or damaged tissues, or may be used to grow new organs.

Scaffolds

Cells are often implanted or 'seeded' into an artificial structure capable of supporting three-dimensional tissue formation. These structures, typically called scaffolds, are often critical, both ex vivo as well as in vivo, to recapitulating the in vivo milieu and allowing cells to influence their own microenvironments. Scaffolds usually serve at least one of the following purposes:

- Allow cell attachment and migration
- Deliver and retain cells and biochemical factors
- Enable diffusion of vital cell nutrients and expressed products
- Exert certain mechanical and biological influences to modify the behaviour of the cell phase

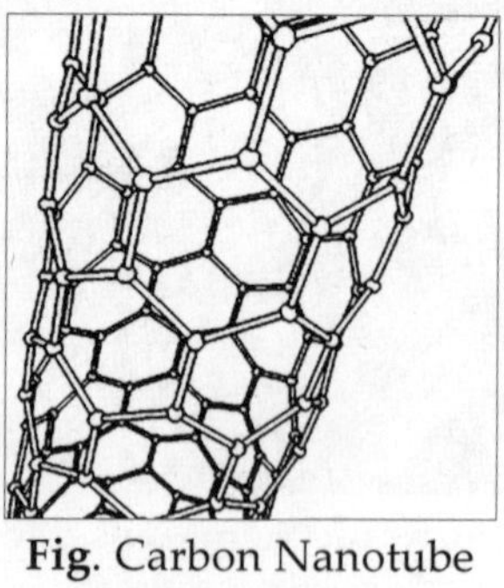

Fig. Carbon Nanotube

Carbon nanotubes are among the numerous candidates for tissue engineering scaffolds since they are biocompatible, resistant to biodegradation and can be functionalized with biomolecules. However, the possibility of toxicity with non-biodegradable nano-materials is not fully understood. To achieve the goal of tissue reconstruction, scaffolds must meet some specific requirements. A high porosity and an adequate pore size are necessary to facilitate cell seeding and diffusion throughout the whole structure of both cells and nutrients. Biodegradability is often an essential factor since scaffolds should preferably be absorbed by the surrounding tissues without the necessity of a surgical removal.

The rate at which degradation occurs has to coincide as much as possible with the rate of tissue formation: this means that while cells are fabricating their own natural matrix structure around themselves, the scaffold is able to provide structural integrity within the body and eventually it will break down leaving the neotissue, newly formed tissue which will take over the mechanical load. Injectability is also important for clinical uses.

Materials

Many different materials (natural and synthetic, biodegradable and permanent) have been investigated. Most of these materials have been known in the medical field before the advent of tissue engineering as a research topic, being already employed as bioresorbable sutures. Examples of these materials are collagen or some linear aliphatic polyesters.

New biomaterials have been engineered to have ideal properties and functional customization: injectability, synthetic manufacture, biocompatibility, non-immunogenicity, transparency, nano-scale fibers, low concentration, resorption rates, etc. PuraMatrix, originating from the MIT labs of Zhang, Rich, Grodzinsky and Langer is one of these new biomimetic scaffold families which has now been commercialized and is impacting clinical tissue engineering.

A commonly used synthetic material is PLA - polylactic acid. This is a polyester which degrades within the human body to form lactic acid, a naturally occurring chemical which is easily removed from the body. Similar materials are polyglycolic acid (PGA) and polycaprolactone (PCL): their degradation mechanism is similar to that of PLA, but they exhibit respectively a faster and a slower rate of degradation compared to PLA.Scaffolds may also be constructed from natural materials: in particular different derivatives of the extracellular matrix have been studied to evaluate their ability to support cell growth.

Proteic materials, such as collagen or fibrin, and polysaccharidic materials, like chitosan or glycosaminoglycans (GAGs), have all proved suitable in terms of cell compatibility, but some issues with potential immunogenicity still remains.

Among GAGs hyaluronic acid, possibly in combination with cross linking agents (e.g. glutaraldehyde, water soluble carbodiimide, etc...), is one of the possible choices as scaffold material. Functionalized groups of scaffolds may be useful in the delivery of small molecules (drugs) to specific tissues.

Synthesis

A number of different methods has been described in literature for preparing porous structures to be employed as tissue engineering scaffolds. Each of these techniques presents its own advantages, but none is devoid of drawbacks.

- *Nanofiber Self-Assembly:* Molecular self-assembly is one of the few methods to create biomaterials with properties similar in scale and chemistry to that of the natural in vivo extracellular matrix (ECM). Moreover, these hydrogel scaffolds have shown superior in vivo toxicology and biocompatibility compared with traditional macroscaffolds and animal-derived materials.
- *Textile technologies:* these techniques include all the approaches that have been successfully employed for the preparation of non-woven meshes of different polymers.

  In particular non-woven polyglycolide structures have been tested for tissue engineering applications: such fibrous structures have been found useful to grow different types of cells. The principal drawbacks are related to the difficulties of obtaining high porosity and regular pore size.
- *Solvent Casting & Particulate Leaching (SCPL):* this approach allows the preparation of porous structures with regular porosity, but with a limited thickness. First the polymer is dissolved into a suitable organic solvent (e.g. polylactic acid could be dissolved into dichloromethane), then the solution is cast into a mold filled with porogen particles. Such porogen can be an inorganic salt like sodium chloride, crystals of saccharose, gelatin spheres or paraffin spheres. The size of the porogen particles will affect the size of the scaffold pores, while the polymer to porogen ratio is directly correlated to the amount of porosity of the final structure.

  After the polymer solution has been cast the solvent is allowed to fully evaporate, then the composite structure in the mold is immersed in a bath of a liquid suitable for dissolving the porogen: water in case of sodium chloride, saccharose and gelatin or an aliphatic solvent like hexane for paraffin. Once the porogen has been fully dissolved a porous structure is obtained.

  Other than the small thickness range that can be obtained, another drawback of SCPL lies in its use of organic solvents which must be fully removed to avoid any possible damage to the cells seeded on the scaffold.

- *Gas Foaming*: to overcome the necessity to use organic solvents and solid porogens a technique using gas as a porogen has been developed. First disc shaped structures made of the desired polymer are prepared by means of compression molding using a heated mold. The discs are then placed in a chamber where are exposed to high pressure CO_2 for several days. The pressure inside the chamber is gradually restored to atmospheric levels.
  During this procedure the pores are formed by the carbon dioxide molecules that abandon the polymer, resulting in a sponge like structure. The main problems related to such a technique are caused by the excessive heat used during compression molding (which prohibits the incorporation of any tempe-rature labile material into the polymer matrix) and by the fact that the pores do not form an interconnected structure.
- *Emulsification/Freeze-drying:* this technique does not require the use of a solid porogen like SCPL. First a synthetic polymer is dissolved into a suitable solvent (e.g. polylactic acid in dichloromethane) then water is added to the polymeric solution and the two liquids are mixed in order to obtain an emulsion. Before the two phases can separate, the emulsion is cast into a mold and quickly frozen by means of immersion into liquid nitrogen.
  The frozen emulsion is subsequently freeze-dried to remove the dispersed water and the solvent, thus leaving a solidified, porous polymeric structure. While emulsification and freeze-drying allows a faster preparation if compared to SCPL, since it does not require a time consuming leaching step, it still requires the use of solvents, moreover pore size is relatively small and porosity is often irregular. Freeze-drying by itself is also a commonly employed technique for the fabrication of scaffolds. In particular it is used to prepare collagen sponges: collagen is dissolved into acidic solutions of acetic acid or hydrochloric acid that are cast into a mold, frozen with liquid nitrogen then lyophilized.
- *Thermally Induced Phase Separation (TIPS):* similar to the previous technique, this phase separation procedure requires the use of a solvent with a low melting point that is easy to sublime. For example dioxane could be used to dissolve polylactic acid, then phase separation is induced through the addition of a small quantity of water: a polymer-rich and a polymer-poor phase are formed. Following cooling below the solvent melting point and some days of vacuum-drying to sublime the solvent a porous scaffold is obtained. Liquid-liquid phase separation presents the same drawbacks of emulsification/freeze-drying.
- *CAD/CAM Technologies:* since most of the approaches are limited

when it comes to the control of porosity and pore size, computer assisted design and manufacturing techniques have been introduced to tissue engineering. First a three-dimensional structure is designed using CAD software, then the scaffold is realized by using ink-jet printing of polymer powders or through Fused Deposition Modeling of a polymer melt.

Assembly Methods

One of the continuing, persistent problems with tissue engineering is mass transport limitations. Engineered tissues generally lack an initial blood supply, thus making it difficult for any implanted cells to obtain sufficient oxygen and nutrients to survive, and/or function properly.

Self-assembly may play an important role here, both from the perspective of encapsulating cells and proteins, as well as creating scaffolds on the right physical scale for engineered tissue constructs and cellular ingrowth.

It might be possible to print organs, or possibly entire organisms. A recent innovative method of construction uses an ink-jet mechanism to print precise layers of cells in a matrix of thermoreversable gel. Endothelial cells, the cells that line blood vessels, have been printed in a set of stacked rings. When incubated, these fused into a tube.

Tissue Culture

In many cases, creation of functional tissues and biological structures in vitro requires extensive culturing to promote survival, growth and inducement of functionality. In general, the basic requirements of cells must be maintained in culture, which include oxygen, pH, humidity, temperature, nutrients and osmotic pressure maintenance.

Tissue engineered cultures also present additional problems in maintaining culture conditions. In standard cell culture, diffusion is often the sole means of nutrient and metabolite transport. However, as a culture becomes larger and more complex, such as the case with engineered organs and whole tissues, other mechanisms must be employed to maintain the culture.

Another issue with tissue culture is introducing the proper factors or stimuli required to induce functionality. In many cases, simple maintenance culture is not sufficient. Growth factors, hormones, specific metabolites or nutrients, chemical and physical stimuli are sometimes required. For example, certain cells respond to changes in oxygen tension as part of their normal development, such as chondrocytes, which must adapt to low oxygen conditions or hypoxia during skeletal development. Others, such as endothelial cells, respond to shear stress from fluid flow, which is encountered in blood vessels.

TISSUE ENGINEERING IN CARDIOVASCULAR SURGERY

For some years, one branch of experimental research at the DHZB has been the fascinating area of the artificial production of complex tissues, known as tissue engineering.

ENGINEERING OF HEART VALVE PROSTHESES USING AUTOLOGOUS CELLS

For heart valve replacement, at present mechanical and biological prostheses are used. These function well but consist of material that is foreign to the body and have the disadvantage that they cannot grow in size. In children and teenagers this is a serious disadvantage as they "grow out of" the replacement valve and need several corrective operations. A potential solution to this problem may lie in the tissue engineering of replacement cardiovascular tissue.

The principle of tissue engineering is to use autologous cells (cells from the patient's own body) to produce viable, functioning replacement tissue that would become integrated into the surrounding tissue, grow with it and form highly durable valves.

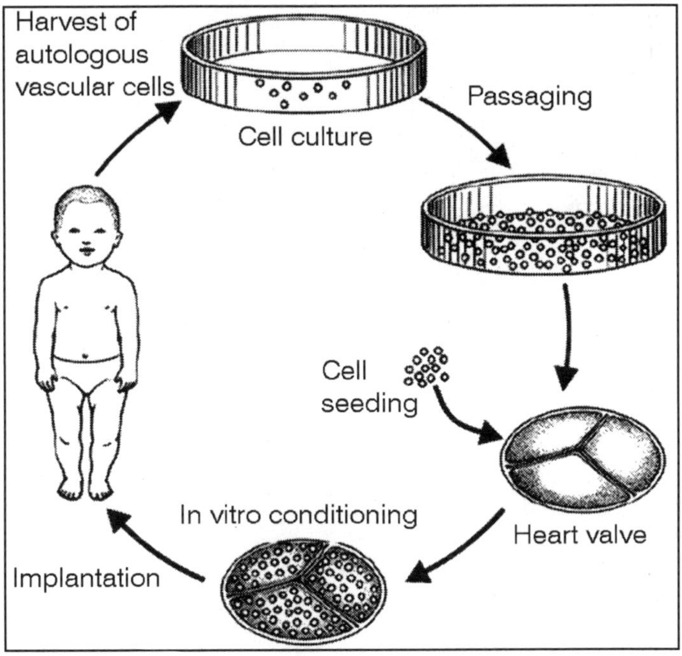

Fig. The principle of tissue engineering

The constructed valve fabricated from human cells and scaffold material, for example a biodegradable polymer, must be produced under optimal in vitro conditions so that it can be later be transplanted into the heart of the patient from whom the cells were taken. A particular focus of research of the DHZB laboratory is the identification of a source of suitable cells. Since it is not possible to use cells from heart valves themselves, cells must be utilized that are as similar as possible to heart valve cells. In the past few years human vascular umbilical cord cells have been evaluated as a source of autologous cells for cardiovascular tissue engineering.

The umbilical cord is normally discarded as superfluous tissue after birth. It has several long blood vessels and can be used to harvest a very large number of vascular cells without the patient undergoing any harvesting procedure. These cells can be stored in liquid nitrogen until required.

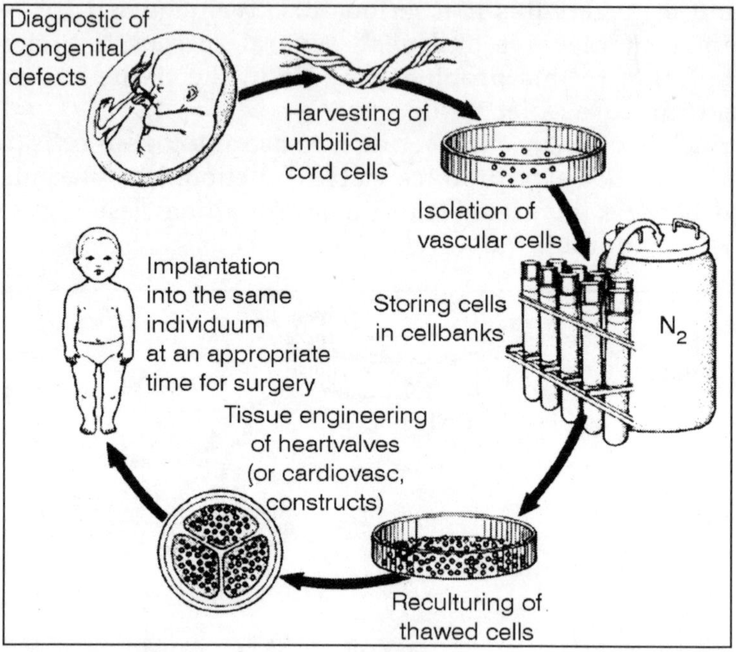

Fig. Cryopreservation of Vascular Umbilical Cord Cells

In our laboratory the vascular umbilical cord cells are characterized using immunohistological and molecular biological methods. The main target group for the possible use of these cells is newborn children with congenital heart valve defects.

In older children and teenagers other suitable cells (mesenchymal precursor cells) are to be isolated from the patients' own bone marrow and used to engineer heart valves. A further focus of our research laboratory is the development of cell culture systems and "bioreactors" in which the heart valve construct is conditioned by mechanical means. By rotating the reactor, for example, it is possible for the cells to be evenly distributed on the scaffold material.

Fig. Cell-seeding device: (1) rotating machine; (2) base frame with rubber rollers; (3) acrylic bowl with balance weight; (4) cell suspension inlet; (5) lid of the bowl; (6) cylindrical cell-seeding chamber

These are the preconditions for the formation of tissue-like united cell structures on a complex three-dimensional scaffold like that for a heart valve. Influenced by pressure and flow in the bioreactor, the cells align in a particular direction.

Thus the formation of extracellular matrix proteins is encouraged, which is very important for the development of cell tissue.

Fig. Bioreactor test

In parallel with this research, the in vitro conditions for fabricating tissue engineered valves are constantly being modified and optimized. The aim is to provide the cells with the greatest possible nutrient supply so that the growth process can be accelerated. In future, viable human heart valves consisting of autologous cells, with the ability to grow, are to be applied in cardiac surgery.

PARTICULATE BIOCERAMICS FOR COMPOSITE SCAFFOLDS

Particulate materials of hydroxyapatite (HA), tricalcium phosphate (TCP) and a few bioactive glasses can be used in composite scaffolds for achieving osteoconductivity. In one of our early investigations, plasma-sprayed HA (psHA) particles were produced and used, as particulate psHA was found to be more bioactive than highly crystalline HA particles due to the amorphous component in psHA particles.

These spherical psHA particles were in the micrometer size range. In our recent investigations, carbonated HA (CHA) particles were synthesized through a nanoemulsion process. CHA is also more soluble and bioactive than highly crystalline HA due to the CO_3 2- substitution in HA lattice. Furthermore, the spherical CHA particles were in the nanometer size range and are thus more suitable for incorporation into polymers to form composite scaffolds.

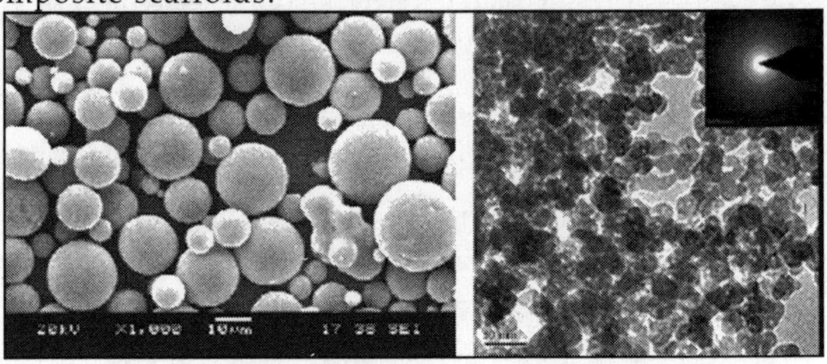

(a) ps HA Microspheres (b) CHA Nanospheres

Fig. Particulate HA for Composite Scaffolds

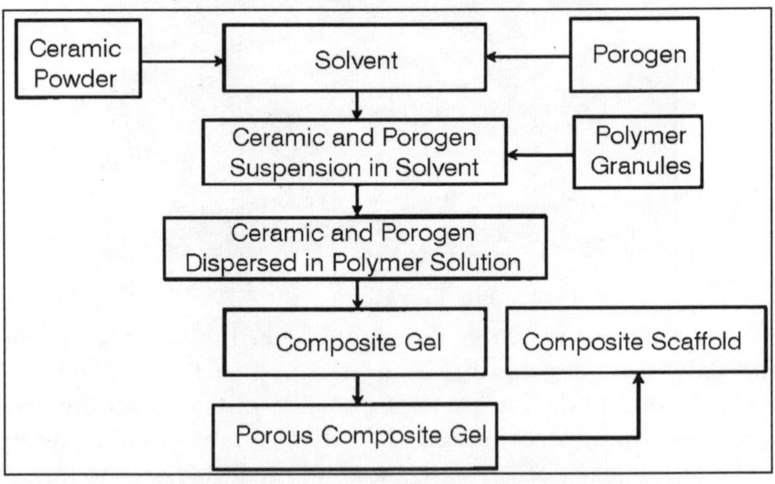

Fig. Manufacture of Bioactve Composite Scaffolds

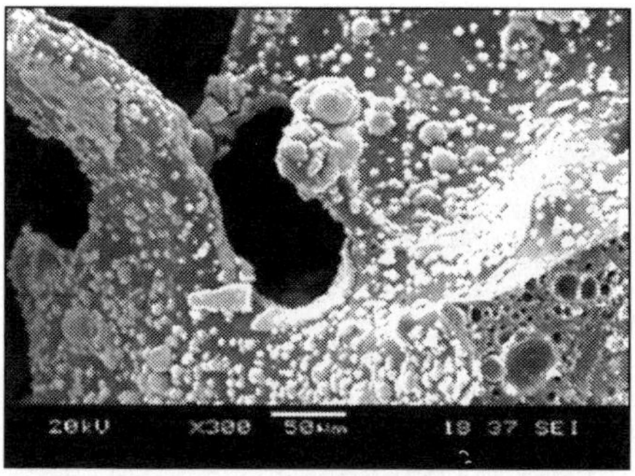

Fig. Formation of bone-like apatite on a ps HA/PLLA Composite Scaffold Through Immersion in Simulated body Fluid

INCORPORATING BIOCERAMIC PARTICLES IN THE SCAFFOLDS

In our early investigations, psHA particles were incorporated in a poly(L-lactic acid) (PLLA) to form osteoconductive composite scaffolds. In this process, freeze-drying was used at the final stage to produce the porous structure. Just as in the production of polymer scaffolds, for composite scaffolds, factors such as polymer solution concentration, porogen type and size, freeze-drying parameters, etc.

play very important roles in forming the scaffolds of desired porous structures (pore geometry, pore size and size distribution, pore interconnectivity, thickness of pore walls, etc.) and hence the mechanical performance. It was shown through in vitro experiments that the composite scaffolds possessed bioactivity due to the incorporation of psHA particles.

In a recent study, it was shown that using an emulsion freeze-drying process, composite scaffolds based on polyhydroxybutyrate (PHB) and polyhydroxybutyrate-co-valerate (PHBV) natural polymers could be produced.

In these scaffolds, the CHA nanospheres were incorporated. Through this process, highly porous structure with a controlled pore morphologies could be achieved.

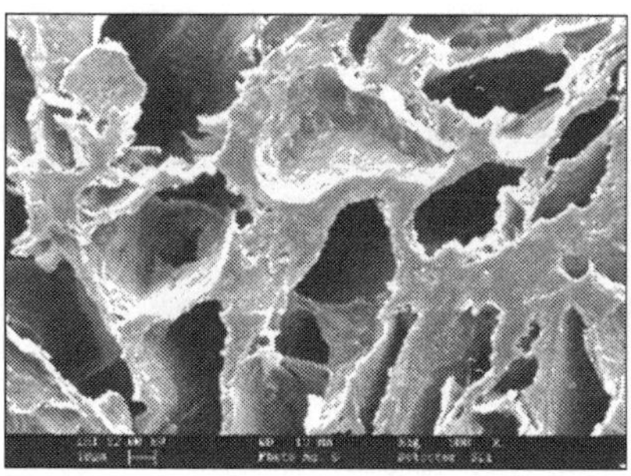

Fig. CHA/PHBV Composite Scaffold Produced through an Emulsion Freeze-Drying Process

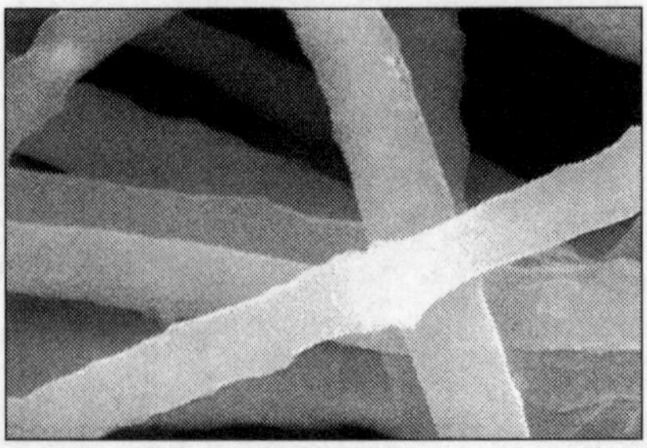

Fig. CHA/PHBV Composite Nanofibers Produced through Electrospinning

In recent years, biodegradable fibers which have diameters in the nanometer range have attracted great attention, as it is considered that the proper in vivo phenotype cannot be consistently achieved if cells are presented with fibers with diameters equal to or greater than the cell size. Among technologies for making polymer fibers of nanometer dimensions, electrospinning is the most appealing and the process is relatively simple. A study was recently conducted on electrospinning of PHB and PHBV polymers and their composites. It was found that electrospinning parameters (polymer solution concentration, solution feeding rate, injection needle diameter, electrical voltage, and working distance) all had significant effects on the resultant fibers. With careful experimentation, CHA/PHBV composite nanofibers containing CHA nanospheres could be produced. The incorporation of CHA nanospheres was confirmed through EDX analyses of the fibers. Rapid prototyping (RP) technologies have become increasingly popular for producing tissue engineering scaffolds, as there are a few distinctive advantages of using RP technologies. One of our efforts is to use the selective laser sintering (SLS) technology to manufacture CHA/PLA composite scaffolds. In the SLS route, there are two major issues:

- Obtaining powders of appropriate characteristics for sintering,
- Minimizing the use of sintering powder.

The latter issue could be solved by building and installing a miniature sintering platform in a commercial SLS machine. The former issue required the production of composite powders of suitable characteristics. Through a microemulsion process, both PLA and CHA/PLA composite powders were made.

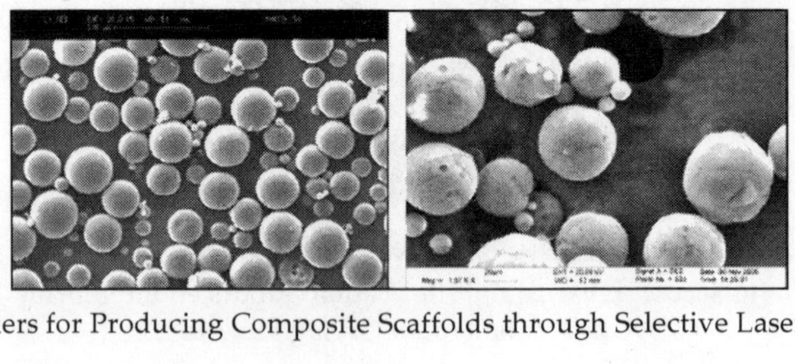

Fig. Powders for Producing Composite Scaffolds through Selective Laser Sintering

Coating the Polymer Scaffold with an Apatite Layer

As stated earlier, scaffolds made purely of biodegradable polymers such as PLA and PLGA are not osteoconductive. To impart bioactivity on the polymer scaffolds and also strengthen the scaffolds, apatite coatings could be formed on pores surfaces of the scaffolds before they are put in use for tissue engineering. For polymers, forming apatite coatings on their surfaces can be achieved through biomimetic processes. However, as polymers such as PLA degrades fast in the aqueous environment, normal biomimetic processes that have been developed for coating apatite on metals or ceramics are not suitable for these polymers. Therefore, an accelerated biomimetic process was developed to coat PLLA scaffolds with a thin layer of apatite.

Apatite was found to form on PGA meshes and PLLA scaffolds, and with a dynamic environment, biomimetic apatite coatings could be produced on pores inside the scaffolds. Further investigations revealed that a composite coating consisting of nano-sized apatite crystals and collagen which mimics bone structure could also be produced on PLLA scaffolds. It was found that the apatite coating and apatite/collagen composite coating could improve interactions between osteoblasts and PLLA scaffolds and that the composite coating was more effective than apatite coating in improving such interactions. Osteoblast-like cells could attach and proliferate on the apatite/collagen coating formed on PLLA via the biomimetic process.

Fig. Bone-like apatite coating formed on the pore surface of a PLLA scaffold through an accelerate biomimetic process

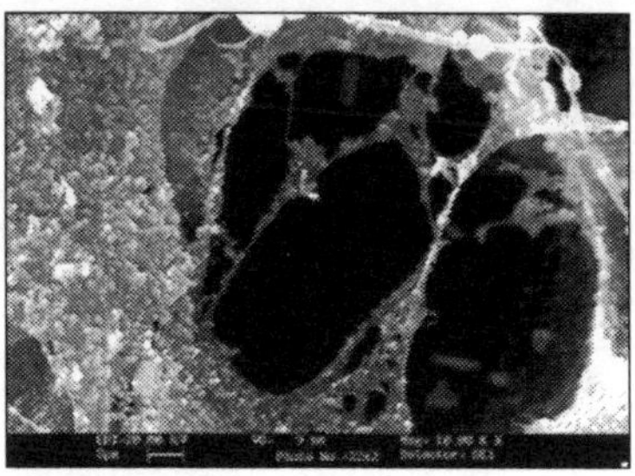

Fig. Apatite/Collagen Composite Coating Formed on the Pore Surface of a PLLA Scaffold through an Accelerate Biomimetic Process

Fig. Osteoblast-Like Cells Attaching to the Apatite/Collagen Composite Coating

Currently, there are a number of strategies for developing bone tissue engineering scaffolds. The composite strategy provides a means for achieving stronger and bioactive scaffolds as compared to conventional polymer scaffolds. The two approaches in the composite route result in scaffolds of different, distinctive characteristics. Bioceramics obviously have an important role to play in constructing osteoconductive tissue engineering scaffolds. Systematic studies are required for the manufacture, characterisation, and in vitro and in vivo assessments of bioceramic-polymer composite scaffolds.

Quite often, the scaffold manufacturing technology is selected together with considerations of choosing appropriate constituting materials of the composite. Osteoconductive composite scaffolds appear to have great potential for bone tissue engineering.

2

Precursors of Tissue Engineering and Technology

Tissue engineering represents the confluence of a complex array of pre-existing lines of work from three quite different domains: the worlds of clinical medicine, engineering, and science. This section offers a brief overview of the range and character of these different contributing elements, to provide a context for the discussion in later sections of intellectual and institutional developments in TE.

ENGINEERING AND TECHNOLOGY OF TISSUE

Engineering and technology are not static bodies of knowledge and practice, but dynamic processes characterized by constant change. In supporting fundamental research in engineering, the National Science Foundation is necessarily planting the seeds of continued change, for the knowledge gained is itself the fuel for this ongoing transformation. But new knowledge does not affect solely the particulars of specific technologies and branches of engineering.

More fundamentally, it may transform our understanding of knowledge itself, of its organization and of the means of creating and utilizing it most effectively, pointing to new ways of thinking about engineering challenges and, ultimately to the emergence of new areas of research and practice that in time come to be recognized as distinct fields of activity. Thus, in fulfilling its mission to preserve and enhance the vitality of engineering in service of the nation, NSF must constantly be alert not only to opportunities for continuing advances within established lines of inquiry, but also to the possibility that entirely new directions for research may offer compelling benefits for engineering and for society – indeed, that they may be essential to continued progress.

In practice, to what extent has NSF served as a catalyst for the timely emergence of productive new domains of engineering research? What are the specific mechanisms by which it has done so? They are difficult questions to answer, for the innovation system of which engineering research is a part – and in which NSF plays an important role – is actually a complex ecosystem characterized by noisy signals traversing tangled pathways of influence and feedback among the system components. Gathering the data that may enable one to make sense of the system is challenging, both because the system is only imperfectly "instrumented" with routine measures of its behaviour, and because ad hoc observations are costly both to the system and to its observers. Finally, the complexity of the pathways of influence that govern the system can make attribution of causality problematic, and the idiosyncratic variety of the particulars in different domains of engineering and innovation make generalization perilous.

This report examines the emergence of the research field of tissue engineering (TE), focusing on developments in the United States. Its purpose is to document the evolution of tissue engineering into a distinct area recognized as such by scholars, to document NSF's involvement through its Directorate for Engineering, and to evaluate the significance of NSF's role in the broader context of the field's evolution. As a topic of study, tissue engineering is interesting for several reasons.

First is the field's inherent human interest. The ultimate goal of the field is to develop powerful new therapies – "biological substitutes" – for structural and functional disorders of human health that have proven difficult or impossible to address successfully with the existing tools of medicine. This goal is made all the more provocative by its science-fiction vision of man-made, living "replacement parts" for the human body, and more compelling by its framing in terms of the lives lost through the ultimately irremediable shortage of transplantable organs.

Of special interest to research managers is the unusual breadth and depth of TE's interdisciplinarity: the knowledge needed to meet the technical challenges posed by TE spans many subdisciplines not only of engineering, but also of science and of clinical medicine. To reach their ultimate goal – successful therapies – tissue engineers must integrate many very different kinds of knowledge and ways of thinking. Finally, from the perspective of an evaluator, the study of an emerging, strongly interdisciplinary field – a complex, ill-defined, moving target – represents a challenge of the first order, demanding ingenuity and flexibility as much as narrowly-defined methodological rigor, in pursuit of the real prize: qualitative insight that is well-founded in empirical observation and credible in its logic.

Where does one begin such a study? In the contemporaneous historical analysis of an emerging, strongly cross-disciplinary field, it can be difficult to specify the object or scope of the inquiry precisely at the outset. Almost by definition, the conceptual frameworks that define the substantive extent of the field remain fluid, while the field's institutional infrastructure is typically both limited and not fully apparent to a superficial review. The investigation will often take on a "bootstrap" character, in which part of the task of the inquiry is to delimit its own scope through analysis of the evidence that is gathered.

Within a biomedical context, the term "tissue engineering" points naturally to a central concept that is simple, powerful and intuitive: the creation of living tissues for therapeutic purposes. It is not obvious, however, how far this concept should extend. What counts as a "tissue"? (For example, are encapsulated pancreatic cells that function autonomously a tissue? Is blood a tissue? Are unorganized stem cells a tissue?) Must the "engineering" be done in vitro, or is induction of tissue growth in situ in a living organism also "tissue engineering"?

Must the engineered product be implanted, or is improvement of the function of extracorporeal devices through the introduction of living cells to be considered "tissue engineering"? Must the product be directly therapeutic, or is the use of complex, cell-based products in diagnostic applications also a type of "tissue engineering"? To what extent or under what circumstances should fundamental research on underlying concepts or enabling technologies be considered tissue engineering?

Thus, the present study began not with a precise definition of tissue engineering, but rather with a set of plausible entry points for inquiry: the names of a few key researchers widely considered synonymous with the field, a handful of review papers, a list of NSF awards in support of TE research and ancillary activities – and the term "tissue engineering" itself, which could be presented as a filter to search engines used to scan bibliographic and research funding databases as well as the Internet. Following a chain of referrals, citations and links from these initial sources, the investigation accumulated many alternative definitions for the term, as well as a growing list of suggestions by expert interviewees of domains of research activity that they believed could or should be considered part of the field.

A central and ongoing task of the study was to analyse the character and overall coherence of the scope of activity delineated in this way. Specific findings in this respect are presented in appropriate context in the later sections of this report. However, it is worth noting here that no simple boundary-setting rule could be identified that maps cleanly to the scope of activity of researchers who think of themselves, or who are thought of by others, as being tissue engineers. In the end, our own concept of tissue engineering remained a pragmatic and operational one. In particular, because part of the charge for this project was to understand the role of NSF in shaping the field, and because funding agencies act through support of people and their activities, final judgment of what to include in the story was shaped as much by the sociological structure of the field – the people and institutions involved – as by its intellectual or substantive structure.

ROOTS IN CLINICAL MEDICINE

Perhaps the most obvious precursors to TE lie in the clinical domain. These are best understood as specific examples of general problem-solving strategies employed by physicians. Consider, for example, the basic dilemma faced by a surgeon. While the removal of organs or body structures that are damaged beyond repair by disease or trauma can be life-saving, the patient must cope with the functional effects of tissue loss and, in some cases, the psychological impacts of disfigurement. And for those vital organs whose complete removal is incompatible with life, the surgeon's hard-earned skill is, by itself, of no avail: the procedure cannot be done unless there is some way of replacing or reconstituting essential functions.

The resourceful application of virtuosic craft through the tools and materials at hand to meet the distinctive needs of each patient is central to the ethos of surgery. Thus, surgeons have sought to reconstruct anatomic structures using the patient's own tissues as raw material; they have pressed artificial materials into service as prostheses; and, most spectacularly, they have brought patients back from the brink of death by transplanting an ever-wider range of vital organs – primarily living organs, but in a few cases, with only very limited success to date, prototype artificial organs as well.

However, with experience, surgeons have come to understand in detail not only the benefits of such measures, but their limitations as well. Anatomic reconstruction using the patient's own tissues can cause substantial morbidity at the donor site; the improvised structures are usually functionally inferior to the natural organs they replace, and less durable as well. Poor compatibility between artificial materials and mechanical systems and the internal environment and physiologic requirements of the human body can lead to dysfunctional interactions and new failure modes. Transplantation of living organs brings with it profound immunologic complications, and the number of patients who can be treated in this way will always be severely constrained by the limited supply of organs suitable for use.

For a surgeon, then, the development of engineered tissues is a logical next step in the ongoing effort to improve the match between the surgeon's various reparative and reconstructive contrivances and the requirements of human anatomy and physiology. Physicians in a range of internal medicine specialties as well have found themselves impelled to explore clinical solutions that incorporate living cells. Generally, internists seek to identify therapies that can repair or reconstitute physiologic functions with sufficient effectiveness to enable patients to avoid surgery.

Often these therapies are pharmacologic – physicians may use small molecules, or, increasingly, complex, genetically-engineered biological macromolecules, to replace critical molecular species that are in short supply within the body, to counter the effects of molecules that are in harmful oversupply, or to intervene in more subtle ways in regulatory pathways that control critical functions. Other therapies rely on physico-chemical effects – sometimes implemented through external artificial devices – to replace critical filtering functions that maintain electrolyte balance and remove metabolic wastes from the body. Yet the metabolic functions of human organs are so complex and interrelated that simple pharmacologic or physico-chemical approaches, while they can be life-saving, nevertheless often constitute highly imperfect solutions whose limitations become increasingly apparent in chronic use. In such situations, the notion of leveraging the kinds of complex, integrated physiologic functions that are accessible only at the level of intact cells or tissues becomes compelling.

Examples

Landmark conceptual developments prerequisite for a concept of tissue engineering have emerged over a period of decades within the problem-solving traditions of clinical medicine.

In some clinical domains, physicians and non-clinician researchers reached the stage of incorporating living cells into prototype tissue-engineered clinical solutions years before the emergence of a generalized concept of tissue engineering.

Vascular grafts

The earliest explorations by surgeons of the possibility of transplanting blood vessels took place more than a century ago; the renowned surgical researcher Alexis Carrel was awarded the 1912 Nobel prize in physiology or medicine for his demonstration of successful techniques for the anastomosis of blood vessels and the extension of these techniques from the transplantation of vessels to the transplantation of entire solid organs. Over the succeeding decades, experimental use of rigid glass and metal tubes as vascular grafts yielded disappointing results. In the early 1950s, Voorhees demonstrated the first use of tubes of synthetic fabric as arterial prostheses.

With the expanded use of a range of synthetic grafts in clinical practice and ongoing research into the characteristics of a range of alternative materials, surgeons and biomaterials researchers gained a deeper appreciation of thrombogenesis and other problems arising from the interaction between synthetic materials and the blood and perigraft tissues with which they came in contact.

The concept of a resorbable vascular graft was introduced in the 1960s, and the first fully-resorbable graft was reported in 1979. Improvement in the healing process of Dacron vascular grafts via pre-seeding with endothelial cells was reported in 1978. Finally, the first attempt to create entirely biologic vascular structures in vitro, using collagen and cultured vascular cells, was reported in 1982.

Skin grafts

For centuries, physicians have attempted to cover severe wounds with grafts from a variety of sources, including cadavers and living humans. By the early decades of the 20th century, researchers were investigating the immunologic basis for rejection of skin allografts, though there was no apparent progress toward any practical solution.

The marked increase during World War II in the number of burn victims for whom a skin allograft was not feasible provided a renewed impetus for research on skin replacement. During this era, the distinguished immunologist Peter Medawar made important contributions both to further progress in the understanding of the immunology of graft rejection, and to the in vitro culture of epithelial cells drawn from a patient. By 1953, Billingham and Reynolds demonstrated in animal models that the products of a brief culture of epidermal cells could be applied to a graft bed to reconstitute an epidermis.

Despite these early successes, more efficient means of cultivation were needed to provide enough cells to sustain transplantation. Overgrowth of certain cell types, such as fibroblasts, suggested that existing culture techniques would not be effective in producing large quantities of cells. Research in the 1960s and 70s identified growth factors that could be added to culture medium to induce greater proliferation of epidermal cells. Starting in the mid-1970s, three research groups working independently at MIT reported a series of milestones in the development of skin replacements.

In 1975, Green and Rheinwald described the co-culture method, a technique for serial cultivation of human epidermal keratinocytes from small biopsy samples.Using the technique, one sample of cells was sufficient to generate enough thick, multilayered skin to resurface the entire body of a burn victim. Some differentiation among epidermal cells to mimic that of the epidermis was also visible. In 1979, Green and colleagues, building on the work of Bellingham and Reynolds, demonstrated that cultured cells can be grown in sheets in a petri dish and transferred intact, rather than as disaggregated cells, to a graft wound bed.

That same year, Bell and colleagues described the use of fibroblasts to condense a hydrated collagen lattice to a tissue-like structure potentially suitable for wound healing. These findings led to the first functional living skin equivalent (LSE) in 1981, consisting of fibroblasts suspended in a collagenglycosaminoglycan matrix. Yannas identified the components of the underlying matrix structure of skin and used this knowledge to develop a dermal regeneration template which, when implanted and seeded with autologous basal cells, stimulated re-growth of functional skin.16 All of these lines of research, continuing into the 1990s, proceeded to the development of commercial products.

Kidney

A number of investigators experimented sporadically with kidney transplantation during the early part of the 20th century, but planned renal transplantation efforts began only in the late 1940s. During the war years in the Netherlands, Kolff developed the first dialysis machine; its design was refined at the Peter Bent Brigham Hospital in Boston and used in patients for the first time in 1948. In turn, the availability of effective short-term dialysis facilitated progress on transplantation, culminating in the successful transplant of a donated kidney from a twin by Murray and colleagues at Brigham and Women'a Hospital in 1954. Subsequent advances in immunosuppression for transplantation and further development of Kolff's dialysis machine made both techniques practical for widespread, routine use, transforming the management of end-stage renal disease.

However, the limited supply of organs suitable for transplantation, combined with recognition of the therapeutic limitations of the dialysis machine, motivated the formulation of the concept of the bioartificial kidney, which would mimic more faithfully the physiologic functions of the kidney and thus avoid the debilitating side effects of chronic dialysis. During the late 1960s and early 1970s, Wolf experimented with combinations of kidney cells with hollow synthetic fibers as conduits for nutrients and waste. Subsequent work on growing liver cells on the outside of hollow fibers by Wolf and independently by Knazek led to the demonstration of hollow-fibre bioreactors. In the mid-1980s Galletti and colleagues furthered development of the bioartificial kidney concept through their research on hollow-fibre bioreactors employing renal epithelial cells.

Pancreas/ islet cells

Although the introduction of insulin more than 70 years ago had a miraculous impact on the lives of diabetes patients, with prolonged survival it became apparent that the highly imperfect glycemic control typical of routine insulin therapy was associated with severe long-term complications for a variety of organ systems. The first insulin pumps appeared in the 1960s, but the sensor and control technology required for a complete "closed loop" system to mimic the adaptive character of physiologic glucose control was not available. The first pancreas transplant, in conjunction with a simultaneous kidney transplant, was performed by Lillehei in 1966.

Lacy reported a method for isolation of intact islets in 1967, and isolated islet cells were first transplanted in 1970, though without a solution to the problem of immune rejection. The use of microencapsulated islets as artificial beta cells was proposed by Chang as early as the mid-1960s. During the 1970s, Chick and colleagues, building on the work of Knazek, developed a "hybrid artificial pancreas" consisting of beta cells cultured on synthetic semipermeable hollow fibers, and demonstrated the ability of this device to restore glucose homeostasis in rats when connected to the circulatory system via shunt. Sun and colleagues reported similar work in the 1970s, followed by studies of implanted microencapsulated islets beginning in 1980.

Investigation of different ways of "packaging" islet cells to provide effective and durable glycemic control continued through the 1980s and beyond.

Liver

The first successful liver transplant was carried out by Starzl in 1967, but in the absence of an adequate supply of transplantable organs, researchers and clinicians have continued to pursue alternative approaches to the replacement of hepatic function. Over more than four decades, clinicians have attempted in a variety of ways to provide extracorporeal support to patients suffering from liver failure. Nonbiological approaches that have been explored include hemodialysis, hemoperfusion over charcoal or resins or immobilized enzymes, plasmapheresis, and plasma exchange.

However, these approaches have met with limited success, presumably because the complex synthetic and metabolic functions of the liver are inadequately replaced by these systems. Work on cell-based therapies and bioartificial systems has reflected the same themes observed in the case of pancreatic islet cells.

Wolf and Munkelt reported in 1975 on a bioreactor containing a rat hepatoma cell line cultured on the surface of semipermeable hollow fibers within a plastic housing. Sutherland and colleagues reported in 1977 on the use of transplanted hepatocytes in the treatment of drug-induced liver failure in rats, and Sun and colleagues in the mid- 1980s explored approaches to microencapsulation of hepatocytes.

Bone and Cartilage

A variety of materials generally perceived as chemically inert, such as various metals and alloys, have been used for many years to replace damaged bone or to provide support for healing bones. With experience, however, it has become clear that non-biologic materials do not remain biologically inert in the environment of the human body, but rather elicit reactions whose intensity is related to a variety of factors such as implantation site, the type of trauma at the time of surgery, and the precise material in use.

Beginning in the 1970's, bioactive materials, such as porous glass and hydroxyapatite ceramic, were examined as alternatives, as they were shown to elicit the formation of normal tissue on their surfaces. Aside from novel biomaterial development, the growth and regenerative capacities inherent in bone have also been intense topics of scientific and clinical study for decades. In a 1945 publication in Nature, Lacroix hypothesized that osteogenin, a substance in bone, was responsible for its growth. Twenty years later in 1965, Marshall Urist proved that there was, indeed, some substance or combination of substances present in demineralized bone, which when transplanted, could induce growth of new bone. Urist's landmark finding encouraged many to investigate the precise factors which trigger bone induction.

During the 1970s and 80s, research demonstrated that the process is mediated by a category of growth factors, termed bone morphogenetic proteins or BMPs, which act in a multistep cascade highly reminiscent of embryonic bone morphogenesis. Reddi and colleagues developed techniques to isolate these proteins from the extracellular matrix of bone. The findings are promising, too, for induction of cartilage as BMPs initially induce a cascade of chondrogenesis and might just as easily be called cartilage morphogenetic proteins. Work on isolation, purification, and proliferation of BMPs continued throughout the 1980s.

EVOLUTION AND EMERGENCE OF TISSUE ENGINEERING

It is unclear who first used the term "tissue engineering" to mean what it does today. Not surprisingly for a coinage that seems so natural in hindsight, a number of the individuals interviewed for this study suggested that the term may have been invented several times independently before it came into sufficiently broad usage that a wide range of researchers can be expected to have encountered it in publications or in discussion. Indeed, the first appearance of the term in print of which the study team is aware – also the earliest revealed through a PubMed search – was an incidental, almost offhand usage in a 1984 publication that described the organization of an endothelium-like membrane on the surface of a long-implanted, synthetic ophthalmic prosthesis.

However, the origin of "tissue engineering" as it is recognized today can be clearly traced to a specific individual. In 1985, Y.C. Fung, a pioneer of the field of biomechanics and of bioengineering more broadly, submitted a proposal to NSF for an Engineering Research Centre to be entitled "Centre for the Engineering of Living Tissues". Fung's concept drew on the traditional definition of "tissue" as a fundamental level of analysis of living organisms, between cells and organs: The study of organs and organ systems has historically been the domain of the physiologist and physician. There is, therefore, a relative wealth of practical information about organs, codified in terms of medical practice. On the other hand, tissues are composed of cells, having specialized internal organelles and, ultimately, chemical constituents. The composition of the cell and its constituents has been dealt with by cell biologist and biochemist.

There are relatively few focused efforts at bridging the gap between these extremes. A clear understanding of phenomena at the tissue level is prerequisite to the engineering of tissues. Fung's proposal was not accepted. Nevertheless, the concept of an engineering approach to the level of biological organization between cells and organs surfaced again at NSF in the spring of 1987, at a panel meeting convened to review proposals to the Bioengineering and Research to Aid the Handicapped (BRAH) Programme within the

Engineering Directorate. Fung was present at this meeting, and is recalled as having volunteered the term "tissue engineering" in the course of a discussion that was seeking to crystallize the concept. Further discussions between the programme directors for the BRAH Programme, which looked at whole organs, and the Directoratefor Engineering's Biotechnology (BIOTECH) Programme, which focused on the cellular level, led to the convening of a special Panel Meeting on Tissue Engineering at NSF on October 28, 1987.

For this meeting, Allen Zelman, a Programme Director for BRAH, prepared a draft definition of tissue engineering: The term "tissue engineering" indicates a new inter-disciplinary initiative which has the goal of growing tissues or organs directly from a single cell taken from an individual. Interestingly, in his "statement of the problem", Zelman pointed to the avoidance of immune rejection through the growth of tissues or organs from a patient's own cells as the key benefit of tissue engineering. Zelman envisioned the development of a large and vigorous new industry producing internal organs, but tempered this vision with the caveat that production of complex internal organs, such as the kidney, "would be considered far too ambitious as a starting point" and that "tissues, being more simple than organs, should be investigated initially".

Subsequent to this meeting, a Forum on Issues, Expectations, and Prospects for Emerging Technology Initiation was held in Washington, DC, under the sponsorship of the Division of Emerging Engineering Technologies within NSF. This forum recommended that tissue engineering be designated as an emerging engineering technology, and that a workshop be held to identify appropriate areas for research in this technology.

This proposed workshop, organized by Zelman, Frederick Heineken, and Duane Bruley –all of NSF—was held at Granlibakken Resort, Lake Tahoe, California, in February 1988. With the exception of a re-publication of the 1984 ophthalmology paper in another ophthalmology journal in 1985, the next known appearance of the term "tissue engineering" in print was in the proceedings of the Granlibakken workshop. A preface to these proceedings defined the term more broadly than did any of the provisional definitions floated up to that point, to encompass a wider range of potential therapeutic interventions that could be enabled by research carried out under this new perspective:

"Tissue Engineering" is the application of principles and methods of engineering and life sciences toward fundamental understanding of structure-function relationships in normal and pathological mammalian tissues and the development of biological substitutes to restore, maintain, or improve tissue function. The basic point of the above definition is that tissue engineering involves the use of living cells plus their extracellular products in development of biological substitutes for replacements as opposed to the use of inert implants. The definition is intended to encompass procedures in which the replacements may consist of cells in suspension, cells implanted on a scaffold such as collagen and cases in which the replacement consists entirely of cells and their extracellular products. At the 1992 UCLA symposium on tissue engineering, Eugene Bell defined tissue engineering in terms of a more specific list of goals:

- Providing cellular prostheses or replacement parts for the human body;
- Providing formed acellular replacement parts capable of inducing regeneration;
- Providing tissue or organ-like model systems populated with cells for basic research and for many applied uses such as the study of disease states using aberrant cells;
- Providing vehicles for delivering engineered cells to the organism;
- Surfacing non-biological devices.

Langer and Vacanti referenced the definition from the Granlibakken proceedings, presenting it in condensed form: Tissue engineering is an interdisciplinary field that applies the principles of engineering and the life sciences toward the development of biological substitutes that restore, maintain, or improve tissue function.

Thus, perhaps the single most cited and influential paper in the field, cites the Granlibakken workshop and builds upon its pioneering definition of the tissue engineering. With this definition as a foundation, they added substance to the notion of common themes underlying a seeming diversity of research by identifying three general strategies for the creation of new tissue – the use of:

- Isolated cells or cell substitutes;
- Tissue-inducing substances; or
- Cells placed on or within matrices.

In the body of the paper, they briefly introduced ongoing efforts across a wide range of organ systems, classified by their embryologic origin – as ectoderm, endoderm, or mesoderm. Finally, they concluded by identifying further common themes, this time in the form of enabling knowledge or technologies of broad significance that should be targets for future research, in the areas of cell biology, cell sourcing and preservation, and materials.

The number of PubMed title/abstract "hits" on the term "tissue engineering" first exceeded 10 in 1994, the year after the Langer/Vacanti review appeared. The number of appearances doubled in 1996, and again in 1998 and 1999, reaching 153 in 1999 and continuing to grow to 214 in 2000. This figure surely understates the extent to which researchers associated the concept "tissue engineering" with their work; once the concept is established, there is no reason for a researcher to call it out explicitly in titles or abstracts unless there is a specific point to be made by doing so. The definitions elaborated during the 1987-93 period provided the basic terms of reference for discussions of tissue engineering through the 1990s.

Tissue engineering researchers seeking to situate their work within a broader context typically cited either the Granlibakken workshop definition or the Langer/Vacanti framing of the field. Even those who did not directly cite these sources offered definitions that reflected some combination of the elements contained in the definitions outlined here, with the exception of Eugene Bell's final point about surfacing non-biological devices, which seems to have been ignored for the most part.

These early definitions left at least two important ambiguities. One involved the role of cells in tissue engineering. In many formulations of the concept, the unique aspect of tissue engineering compared to traditional biomedical engineering or to pharmaceutical development was that its products incorporated living cells.

Bell's 1992 definition, allowing for "acellular replacement parts capable of inducing regeneration", reflected an emerging recognition that physiologic reactions to biomaterials were not necessarily just a nuisance to be suppressed, but might under some circumstances offer a means of inducing useful adaptations in the body. However, this conceptual advance also blurred the distinction between this new field and the studies of acellular biomaterials that had been a mainstay of biomedical engineering and materials science. Langer and Vacanti went one step further, defining the use of "tissue-inducing substances" more generally as one of the strategies of tissue engineering. Including growth factors or other "signaling molecules" within the domain of tissue engineering represented progress toward an integrated understanding of the factors that govern tissue development in vivo, but also broke down the boundaries between tissue engineering and modern pharmaceutical research, which draws increasingly on the latest findings of cellular and molecular biologists.

In the words of tissue engineer Jeffrey Hubbell, "Doing tissue engineering with factors to stimulate cells in the body is really just fancy drug delivery. One is delivering a drug – like a protein, a morphogenic factor – that stimulates cellular responses at a site with the goal of ending up with some overall tissue reconstruction or regeneration at that site." A second ambiguity concerned the role of hybrid devices in tissue engineering, and the related question of whether therapeutic products of tissue engineering were necessarily intended to be implanted into the body. Several lines of research typically referred to as tissue engineering pursue the development of external bioreactors that can replace critical metabolic functions.

From both a conceptual and a historical perspective, this work arguably represents an incremental advance on the dialysis machine. The longterm vision of researchers working on hybrid devices typically extends to more compact, self-contained versions that can be implanted. However, even when miniaturized, current "bioartificial organs" remain more machines than living organs, closer to today's mechanical artificial heart than to the vision of an adaptive biological implant that is seamlessly incorporated into the body's reparative and homeostatic mechanisms. The linkage of TE research with clinical medicine, although inevitable and desirable in view of the goals of the endeavor, nevertheless has also served as something of an obstacle to the sharpening of a definition of tissue engineering as an academic discipline. For example, orthopedic surgeons have been investigating many different kinds of implant that may promote bone regeneration.

From a clinical perspective, what matters is not so much whether the active agent in an implanted matrix consists of stem cells, bone morphogenetic proteins, or gene therapy vectors, but whether it is therapeutically effective. Similarly, a nephrologist may view the dialysis machine, the external biohybrid "artificial kidney" or functional, histocompatible "microrenal" units created via nuclear transplantation as possibilities along a seamless spectrum of therapeutic options rather than in terms of the radically different underlying technologies they represent.

Recent developments in nomenclature reflect this persistent ambiguity in scope and focus. The National Institutes of Health Bioengineering Consortium (BECON) symposium on tissue engineering, held at NIH in June, was entitled "Reparative Medicine: Growing Tissues and Organs". The proceedings of the symposium offer multiple competing – and to some extent conflicting – definitions of reparative medicine, of tissue engineering, and of the relationship between the two.

At one extreme, reparative medicine is defined very broadly as "the replacement, repair, or functional enhancement of tissues and organs", a definition that is not much narrower than all of clinical medicine, although most of the examples cited have a surgical flavour; tissue engineering is viewed as one strategy among many for reparative medicine. At the other, reparative medicine is defined as a synonym for tissue engineering:

Reparative medicine, sometimes referred to as regenerative medicine or tissue engineering, is the regeneration and remodeling of tissue in vivo for the purpose of repairing, replacing, maintaining or enhancing organ function, and the engineering and growing of functional tissue substitutes in vitro for implantation in vivo as a biological substitute for damaged or diseased tissues and organs.

This definition of tissue engineering is noteworthy for excluding extracorporeal bioartifical organs, and indeed, such devices gain only a passing mention in the proceedings, which otherwise cover a very broad scope. The term "regenerative medicine", offered as a further synonym, appears to have been coined by William Haseltine, to capture for promotional purposes his view of the future of medicine. Contrary to the usage at the BECON symposium, Haseltine's conception positions TE as a subset – a "thread" or "phase" – of regenerative medicine, not as a synonym for it, emphasizing the in vitro construction of human organs for implantation, using specialized biocompatible materials, signaling molecules, and adult human cells. In practice, however, there is little to distinguish Haseltine's "regenerative medicine" from other conceptions of tissue engineering. Despite these alternative forms of the term, however, the true sentiment of the field appears to have been captured early on at the NSF meetings of 1987 and 1988. It is this concept of the field, which has been carried on by its leading proponents and remains highly referenced today.

Development of the Field: 1987-2002

The publication record indicates that the volume of research carried out under the rubric of tissue engineering has increased substantially since 1987, and especially since the mid-1990s, though it is difficult to determine this volume exactly because of the challenge inherent in attempting to precisely specify the scope of the field. A convenient proxy for the scope of TE today is arguably the pair of reference volumes Principles of Tissue Engineering and Methods of Tissue Engineering, which cover an impressively broad range of research subtopics and researchers. The scope of research referenced by these volumes overstates to some extent the reach of tissue engineering today, because many of the citations refer to prior art or to adjacent fields from which current lines of research have drawn concepts and methods.

Much of the scope of knowledge represented in these volumes was created through research efforts not originally conceptualized as investigations in tissue engineering, but which have, nevertheless, contributed to the field's emergence. Nevertheless, the growth in tissue engineering proper – defined here as work perceived or designated by its participants as TE – has been substantial. This growth derives from multiple sources:

- New graduates or established researchers who have chosen to enter the field have initiated or expanded work under established research themes.
- Established researchers who have begun to collaborate with tissue engineers, or who have recognized similarities between their own work and that of tissue engineers as awareness of the field has grown, have relabeled existing lines of research as TE. One example is an apparent increase in the propensity of researchers in orthopedic surgery to conceive of their work as tissue engineering.
- The definition of the field undergoes an implicit expansion when adjacent fields report advances that appear to address core challenges in tissue engineering. A prominent example of this phenomenon is the explosion of research on stem cells within the past few years, in the wake of discoveries in the late 1990s related to embryonic stem cells.

Sourcing and cultivation of cells with desired and stably expressed properties has been recognized as a central research challenge for TE since it was defined as a field. In the words of one prominent researcher, however, "prior to the burst of stem cell activity, there would have been surprisingly little to say regarding progress in living cell therapy or knowledge of the conditions that would enable the practical use of cells in tissue engineering beyond skin."

Today, stem cell research is a vigorous and important component of TE, partly through pursuit by researchers who consider themselves tissue engineers, and partly by extension of the concept of TE to incorporate the freshly relevant work of investigators who see their research as situated within other intellectual domains.

The individuals interviewed for this study found it difficult to identify seminal papers, events or specific discoveries or technical advances that could be characterized as having defined the direction or character of the field. Tissue engineering's growth and development might be better described as the result of incremental progress along several originally independent lines of work, rather than the product of a handful of major breakthroughs or discoveries.

Interviews and bibliometric analysis, pointed to two early papers that have played especially important roles in shaping the overall character of the field. While the 1987 Granlibakken conference officially presented and defined the term "tissue engineering", the 1993 Langer/Vacanti review paper in Science introduced the concept of tissue engineering to a wider audience, alerted many researchers who were independently pursuing related work that others shared similar interests within a larger framework, and provided a convenient label for these activities. On the face of it, the work presented in this paper represented a modest advance. Conceptually, it reflected a logical combination of existing approaches – cell-seeding of two-dimensional matrices of biological origin, as in the early work on artificial skin; three-dimensional cell culture on synthetic matrices; and selective cell transplantation, as in the early work on islet cell transplantation.

However, the method of seeding cells on resorbable polymer scaffolds was unique and rapidly became both the most important enabling technology and the most important organizing concept in the field, serving as a common element across lines of research addressing a wide range of therapeutic challenges. As a technique for building tangible objects, it also became a vehicle for enhanced public visibility – if not enhanced public understanding – of the field and its goal of "growing organs".

The scaffolds-and-cell-seeding technique catalyzed a flurry of tinkering on a wide range of tissue and organ systems, overshadowing to some extent the more fundamental efforts proceeding in parallel to develop the underlying knowledge needed to make the products of this technique viable as therapies. Beyond the obvious need for new scaffold materials with properties optimized for specific tissue engineering applications, key knowledge gaps in the late 1980's and early 1990's included, among others:

- Sources of large quantities of cells reliably and controllably expressing desired phenotypes
- Details of the immune response to implanted tissues, and means of controlling it
- The role of chemical and physical signals in morphogenesis and in the in vivo remodeling of implanted tissues
- Means of controlling angiogenesis in order to achieve adequate vascularization of threedimensional tissue constructs
- Design principles to create and optimize bioreactors and bioprocessing techniques for the manufacture of specific tissue-engineered products
- Means of preserving TE products between the point of manufacture and the time of usage
- Methods for characterization and functional assessment of engineered tissues both in vitro and in vivo

Many researchers began – or continued – to pursue these questions in their own work, and to draw relevant insights from developments in research outside of tissue engineering. In 2002, after 15 years since the initial NSF meetings, TE remains a mix of topical foci and research styles, reflecting in part the heterogeneous origins, intellectual traditions, and disciplinary affiliations of the mix of clinicians, engineers and scientists who work in the field.

Although recognition of the importance of gaps in the fundamental knowledge underlying tissue engineering is widespread, many of the individuals interviewed for this study referred to the persistently "Edisonian"character of much of the work in TE, by which they meant a sort of inspired, ad hoc tinkering focused on the solution of specific practical problems in the creation of usable products. Some considered this a positive attribute while others viewed it as a drawback, reflecting a persistent tension between two different strategies for TE. Is it best to invest in fundamental research that will lay strong theoretical and methodological foundations for the long-term productivity of TE, or are clinicallysignificant products sufficiently close to being within reach as to warrant an Edisonian sprint toward their creation?

It might be expected that Edisonian approaches would be most strongly associated with TE research and development efforts in the corporate sector, while the academic sector would be more strongly focused on fundamentals. The former is certainly true, and because the corporate sector has accounted for the great majority of the funds invested in TE, it necessarily follows that the character of corporate R&D has had a substantial impact on the character of the TE enterprise overall. Many of our informants observed that corporate R&D efforts in tissue engineering have had a modest effect on the progress of the field. Corporate R&D has focused on the creation of proprietary intellectual content centered on the challenges of bringing products to market, and less on the solution of broader challenges in science or engineering. Knowledge transfer from industry back to academia has been limited.

However, many respondents suggest that the Edisonian approach remains a powerful force within the academic sector as well. In part, this reflects the natural inclination of some workers in the field, many but not all of these clinicians who bring to their work a strong practical bent. Some observers believe that another influence – a deleterious one – has been the combined effect of a shortage of funding from traditional sources of support for academic research together with the incentives created by the venturecapital- funded boom in biotechnology startup companies during select periods in the 1980s and 1990s, that has induced some researchers to attempt prematurely to "productize" their ideas or research findings. Given the eclectic nature of the field, it is difficult to make judgments as to the level of progress that has been made in the years since 1987. One way of interpreting the significance of the events of 1987 is that they marked the beginning of an attempt by engineers to systematize and formalize the field of tissue engineering.

The principle of rational design is central to the engineering approach. In turn, rational design is made possible by the elucidation of theoretical principles of broad generalizability and by the systematic characterization of available materials and methods in terms of the parameters that comprise these theoretical models. In the tissue engineering context, some of these principles would need to come from engineering – for example, those related to mechanical aspects of tissues, the behaviour of biomaterials, and processes for producing, preserving and distributing TE products.

Others would need to come from biology – for example, the behaviour of cells and of growth factors. Still others would need to come from clinical medicine – for example, principles of physiology and pathophysiology. No matter whether their disciplinary roots have been in medicine, in engineering or in biology, TE researchers have from the earliest days of their involvement recognized that the future success of the field depends heavily on strengthening the base of systematic knowledge underlying TE applications.

Yet it was engineers who first sought to articulate this point clearly and make it the foundation for a formalization of the field. While this principle is sound, however, the development of the field since 1987 reflects little progress toward a systematization of TE through the creation of a foundation of broadly applicable theory or even a well-structured phenomenology.

Although a great deal of new knowledge has been accumulated, deficits in fundamental understanding cataloged in recent reviews are similar in general outline to those recognized in the late 1980s and early 1990s. Researchers today have gained a much more detailed and sophisticated understanding of the specific challenges that must be addressed, however, and some progress has been made in framing research challenges in particular areas of TE in a more systematic way. Perhaps the most important explanation for this slow progress is simply that the rationalization of TE represents an intellectual challenge of enormous magnitude. Construction of replacement tissues and organs, or controlled induction of endogenous reparative capacities to restore tissue structure and function, represent extraordinarily difficult systems engineering problems, and knowledge both of the behaviour of the system components and of the necessary principles of systems integration remains primitive in relation to what is required.

Another factor affecting the rate of progress may be important gaps in the intellectual resources that have been brought to bear on the challenges of tissue engineering, and in the degree of cross-disciplinary integration that has been achieved. The need for cell therapy is well recognized even by the nonscientist, as evidenced by the perceived need for some form of stem cell research. Yet the complex nature of dealing with actual cells themselves to achieve an outcome is still not fully realized. While there is burgeoning information on the genetics front, advances in technology for rapid proteomic analysis, rapidly growing information on factors that effect (sic) cell lineage, identification of transcription factors involved in the development of tissue structure and control of morphogenesis, and advancing preclinical and clinical research on cell implantation, there is a very important need to bring all aspects together.

Engineers and physician scientists have been instrumental in leading the way in academic tissue engineering research, although they desperately need the participation of workers in other disciplines, such as molecular biologists and cell biologists, to fill the important gaps in understanding between them. We must not be naïve. How do we stimulate interest in critical areas such as applied research in cell biology and foster interdisciplinary collaboration? How do we provide academic recognition for being an important part of a significant achievement? Academic laboratories must be given an incentive to work on common goals. New paradigms and metrics must be established both in academics and industry as we delve into complex biological problems that are well beyond a single scientific or engineering discipline.

With respect to its headline goal as well – to create living replacement parts for the human body – the progress of TE has been slow. As with the underlying scientific challenges, the work of the past fifteen years in tissue engineering has served above all to clarify our understanding of how difficult it will be to achieve the full extent of TE's therapeutic vision. In his 1987 draft concept memo, NSF's Allan Zelman identified a list of "types of tissues most likely to bring early success". In Zelman's words, these were:

- Skin: replacement of existing skin damaged from burns, scars, etc.
- Bone: present artificial hips and other joints could be replaced with hips and joints composed primarily from the patients' own tissue

- Blood vessels: arteriovenous shunts for hemodialysis patients and heart bypass patients would benefit greatly
- Cornea: this could eliminate rejection, bring sight to those who reject corneal transplants and as success grows possibly provide an alternative to eye glasses
- Cartilage: providing cartilage replacement for arthritic patients could bring relief from pain to millions
- Nerves: every year thousands of paraplegics are generated and this may be the means to reconnect nervous tissue too damaged for self-repair
- Blood or blood components: production of viral free blood and blood components could justify a great research effort.

Preliminary progress has been made in the development of many of these tissues, though it is understood that much more work is required before "off-the-shelf" products will be available:

Skin

Skin is perhaps the most successful of the tissue engineered therapies, with several products having completed clinical trials, met with FDA approval, and made the transition to market. In 1997, the FDA approved TransCyte, a skin replacement tissue made by Advanced Tissue Sciences, which consists of dermal keratinocytes grown on a biodegradable polymer. TransCyte serves as a temporary wound cover for burns as new tissue forms. Apligraf, manufactured by Organogenesis, utilizes live human skin cells to form a dual layer skin equivalent approved by the FDA to treat diabetic leg and foot ulcers. Recent advances in skin tissue engineering have resulted in the following examples of products in the last 5 or so years:

- *EpiDex*: From Swiss-based Modex Therapeutics for treatment of chronic skin ulcers; EpiDex grafts are grown from hair follicle stem cells.
- *Dermagraft:* Was introduced in 1998 by Advanced Tissue Sciences. Dermagraft is a cryopreserved human fibroblast-derived dermal substitute, composed of fibroblasts, extracellular matrix, and a bioabsorbable scaffold.
- *Integra:* Integra is a two-layered dressing and is completely acellular. The top layer serves as a temporary synthetic epidermis; the layer below serves as a foundation for re-growth of dermal tissue. The underlying layer is made of collagen fibers that act as a lattice through which the body can begin to align cells to recreate its own dermal tissue.
- *Epicel*: Also manufactured by Genzyme Biosurgery, is the only autologous skin graft that can permanently close a burn wound. Epicel was developed based on original research done by Howard Green.

- *Alloderm:* (LifeCell) is a cell-seeded allogenic skin replacement. The product consists of human dermal collagen seeded with allogenic fibroblasts. The material has recently been launched in the US - initially for patients with third degree burns and limited donor-site tissue.
- *Xenoderm:* Another product from LifeCell consists of porcine dermis used as a replacement for burn wounds. LifeCell claims that experimental data shows consistent incorporation of the matrix into the wound bed, low immunogenicity, and re-population with host cells.

Companies have approached the development of skin equivalents from different perspectives: autologous cellular replacements (Genzyme Biosurgery), allogeneic cellular replacements, and completely acellular replacements (Integra). Each of these appear to achieve success as wound coverings. However, scar tissue formation and wound contraction issues remain problematic.

Available products also fail in several ways to mimic the structure and function of native skin. Substitutes have long acted as passive wound covers, lacking certain essential functions/components—including hair follicles, and glands. Development of such enhancements are the focus of current research in living skin equivalents and suggests the use of stem cells as a basis for development of fully differentiated skin equivalents. Choice of matrix support to maintain fibroblasts and keratinocytes is also still being investigated.

Vascular Grafting

Progress to date in the development of tissue engineered vascular grafts has focused on mimicking the three layers of the normal muscular artery, using combinations of live cells, bioresorbable and non-bioresorbable scaffolding constructs. At present, there are no FDA approved live vascular replacement therapies. Several techniques are in pre-clinical trial but face challenges that may prevent their widespread use/application in the near future."

Huynh and colleagues at Organogenesis and Duke University, for example have used porcine intestine as a graft base for seeding of endothelial cells, which will grow and develop into vessel like structures. The use of porcine cells, while important for clinical research, have unknown effects if transplanted into humans. Other sources for graft bases are also being explored, including fibrillar collagen and bovine collagen gels.

However, none of these have produced a vascular substitute with the mechanical properties and strength of native blood vessels. Traditional problems plaguing the field, including clotting and scar tissue formation also persist in cellular replacements and prevent laboratory products from making it to the clinical trial stage. To combat such problems, researchers have attempted to embed the graft materials with antibiotics and antithrombotic coatings with limited success. There is also the need to create a functional nerve supply and capillary network in vitro to support live vascular tissues. Until such challenges are remedied, prosthetic grafts, made of substances like Dacron and polytetrafluoroethylene, will continue to serve as the major therapy.

Kidney

As a highly complex organ, whole kidney replacement organs are far from being a reality. However, progress has been made in development of temporary replacement devices, such as extracorporeal kidney assist devices. Dr. David Humes, Chairman of The Department of Internal Medicine at The University of Michigan, Ann Arbor, has successfully completed in vivo testing of a Renal Tubule Assist Device (RAD) for treating acute renal failure. The only other treatments currently available for acute renal failure are hemofiltration and dialysis. Extracorporeal devices may improve the outcomes of these patients while making treatment much less costly.

Pancreas/Islet cells

Islet cell transplantation techniques have consisted of two major approaches: perfusion devices and microencapsulation. Perfusion devices, though developed as early as 1970, have failed to make it to the clinical trial stage to due long-term biocompatibility issues, membrane breakage, and size limitations (a problem which plagues bioartificial implantable livers as well). Microencapsulation, has also been in existence for several decades. Refinements to this technique over time involved improving the biocompatibility of the encapsulating materials. Some researchers suggest that widespread clinical application of microencapsulation techniques is just around the corner. Several commercial tissue engineering approaches to repair/replace pancreatic function describe the current state of the field:

- Metabolex is developing proprietary technologies for the microencapsulation of insulin-producing tissues using thin, conforming, biocompatible coatings.
- BetaGene is a privately held biotechnology company developing innovative strategies for the detection and treatment of diabetes. This company was formed for the purpose of developing proprietary technology originating at the University of Texas Southwestern Medical Centre. BetaGene retains exclusive license to aspects of this technology including the use of engineered cell lines for the treatment of type I and Type diabetes and the use of these cells for bulk insulin production.
- Circe Biomedical has developed the PancreAssist System, consisting of a single tubular membrane surrounded by insulin-producing islets, which are, in turn, enclosed within a diskshaped housing.

The tubular membrane is porous and permeable to glucose and insulin.

Liver

Several bioartificial liver (BAL) bioreactor designs have been developed in the laboratory to replace liver function. The basic design of a BAL device consists of circulating patient plasma extracoporeally through a bioreactor that houses/maintains liver cells (hepatocytes) sandwiched between artificial plates or capillaries.

Bioreactor materials have either a spherical shape, large surface area, large pores or high porosity, or are hydrophilic and biocompatible.These features can help to achieve the high density cultures of hepatocytes required. However, there is no one material that possess all of these desired properties. Researchers are actively seeking a support matrix that could provide all these properties in order to have a BAL with improved efficiency and effectiveness. Clinical trials of some BAL devices are already underway in the United States and the UK Circe Biomedical currently has the HepatAssist liver device in clinical trial, which is a extracorporeal device consisting of a hollow-fibre bioreactor lined with porcine cells.

Numerous other challenges plague the development of tissue engineered livers:

- Human hepatocytes are limited in supply which make harvest and culture for liver assist devices difficult
- Hepatocytes are extremely difficult to stabilize and maintain in culture and lose their specificity rapidly. Several devices have made it to clinical trial but fail to seek FDA approval based on the lack of stabilization of the cellular component.
- The liver is so complex and varied in its functions that generating a replacement device that performs all these duties is far from a reality.
- Microencapsulation techniques have also been tried, but have been unsuccessful, again, due to the rapid loss of functionality of liver cells once removed from the body.

Bone, Cartilage

Like skin, tissue engineering of bone and cartilage has experienced relative success as compared to other tissue engineered products. Current strategies consist of two major approaches: Transplantation of osteochondral grafts and transplantation of chondrocytes. Cell populations from cultured periosteum have the ability to form new bone and cartilage under the appropriate conditions and with the addition of the appropriate growth factors. Transplantation of osteochondral grafts, however, runs a possible risk of rejection in the recipient.

Current products and strategies include:

- Carticel, by Genzyme Biosurgery of Cambridge, MA, which has received FDA approval to replace damaged knee cartilage. The product uses autologous chondrocytes and grows them in a biodegradable matrix, which is then transplanted in place of the damaged tissue.

- Stryker Biotech of Hopkinton, MA has an FDA approved OP-1 Implant under the Humanitarian Device Exemption (HDE). The OP-1 Implant is now available across the country and is indicated for use as an alternative to patients' own bone in recalcitrant long bone nonunions where an autograft is unfeasible and alternative treatments have failed.
- Arnold Caplan of Case Western Reserve has performed mesenchymal stem cell (MSC) transplants in animals, and is working on similar transplants in humans. MSC's have been found to induce bone and connective tissue growth.
- Antonios Mikos at Rice University has developed an injectable copolymer that hardens quickly in the body and provides a surface to guide severed long bone regeneration.

To our knowledge, no other allogeneic, cell-based organ- or tissue-replacement product is close to market. Autologous efforts remain dominant. Efforts to bring to market tissue-engineered products that address defects in complex metabolic functions or replace vital organs will require more time and effort before reaching success.

Research and development programs on various approaches to the bioartificial pancreas are said to have consumed over $200 million of private sector funds to date, but designs capable of routine success in large animal models are yet unavailable, while encapsulated cell therapy has failed to demonstrate efficacy in phase III clinical trials. Extracorporeal replacement of critical metabolic functions of the kidney and liver has reached the stage of small clinical trials – Phase I (safety) for the former and Phase II (preliminary safety and efficacy) for the latter. In both cases, the devices' mode of operation involves extracorporeal blood circulation comparable to that of a dialysis machine, and both are initially targeted toward treatment of acute, life-threatening metabolic failure. The tissue-engineered replacement heart remains a distant vision.

The beginning, more subtle conceptions of TE have extended its scope to encompass not only the production of replacement parts that embody the necessary structure and function, but the possibility of induction of endogenous reparative capabilities as well. In principle, such induction may be achieved via a variety of approaches, including implantation of cells that express growth factor molecules, implantation of non-living materials (for example, a collagen sponge) containing growth factor molecules, delivery of genes that encode the required growth factor, or by local or systemic infusion of growth factor molecules. The observed physiologic effects of the "skin replacement" products in promoting wound healing suggest that they might also be described as the first "induced repair" products rather than as replacement organs. Acellular "skin replacement" products on the market, such as the INTEGRA dermal regeneration template derived from the work of Yannas, are designed to function in this way. After more than 30 years of research on bone morphogenetic proteins, a BMP product has also recently reached the market – Medtronic's INFUSE bone graft product, incorporating recombinant human bone morphogenetic protein rhBMP-2. Several companies market acellular matrix materials for bulk applications in orthopedic and reconstructive surgery.

In bringing therapeutic products to market, tissue engineers must surmount not only daunting technical challenges, but regulatory and business obstacles as well. The regulatory environment for cell-containing products is complex and still at an early stage in its evolution; it imposes a substantial financial burden on the product development process, directly through the efficacy standards the product must meet and through the cost of funding the trials needed to demonstrate that efficacy, and indirectly through the financial effects of delay in bringing products to market. Finally, for a product to be viable, it must be possible to develop it, achieve regulatory approval, manufacture it, distribute it and market it at a price adequate to yield a positive economic return.

The difficulty of companies like Organogenesis and Advanced Tissue Sciences in recent times also raises concerns around the financial viability of some tissue engineered products. As of this writing, both Organogenesis and Advanced Tissue Sciences are undergoing reorganization under bankruptcy protection, and on trends to date it appears unlikely that revenues from their artificial skin products will ever cover the cost of the capital invested in their development. At the aggregate level, cumulative investment in tissue engineering research has been estimated to exceed $3.5 billion, of which well over 90 per cent has been provided by private sources, with negligible financial return.

Such concerns are well known by individuals in the public and private sectors and will be important considerations in strategy development for building not only clinically viable, but commercially viable products. It is clear from these examples that despite notable contributions and advancements, tissue engineering is still a field in its infancy.

Whether tissue engineering as we know it today will prove to be a powerful general strategy for developing therapeutic products and methods that can meet the dual hurdles of therapeutic efficacy and commercial viability remains to be seen. A strong research effort is underway, however, and advances in other emerging areas of science, such as stem cell research, are likely to make significant contributions toward helping tissue engineering to become a viable field.

THE TISSUE ENGINEERS AS A GROUP

Disciplinary affiliations of tissue engineers are difficult to analyse in a precise quantitative way, because the specific mix of research areas that are included in a department with a given name or are associated with a degree with a given designation, and the number and character of departmental affiliations awarded to faculty, vary from one institution to another in idiosyncratic ways. Nevertheless, overall trends emerge clearly from a rough tally of the available data.

Viewed in terms of both the departments by which their doctoral degrees were awarded and the departments in which those who are in academia now hold faculty appointments, tissue engineers are indeed predominantly engineers.

By training, more than half of the individuals for whom we have specific data are engineers. Approximately one fifth of the tissue engineers in our sample hold medical or dental degrees, often in conjunction with an engineering or science doctorate. In view of the critical importance of cells and of physiology in tissue engineering, the fraction of individuals who hold only a doctorate in biological sciences (for example, biology, biochemistry, or non-clinical medical sciences such as physiology or anatomy) is remarkably small – roughly one out of ten. Chemical engineering is by a wide margin the engineering discipline most frequently encountered in this cohort, followed by biomedical or bioengineering, and by mechanical engineering. A more liberal interpretation of which aspects of contemporary orthopedics and biomechanics research should be considered part of tissue engineering would perhaps have yielded somewhat increased weights for biomedical and mechanical engineering in the sample.

Current academic departmental affiliations of these tissue engineers are also strongly weighted toward engineering. However, here bioengineering or biomedical engineering is the leading disciplinary affiliation by a wide margin, followed by chemical engineering and, in a distant third place, mechanical engineering. This pattern may reflect the relatively recent emergence of biomedical engineering as a discipline, and the migration of faculty – especially recent trainees taking their first faculty appointments – from some of the classical engineering disciplines to newer and often growing departments of biomedical engineering.

Biological science affiliations are few, and are weighted toward basic medical science departments. Clinical and clinical science departmental affiliations are strongly weighted toward surgery and surgical specialties, notably orthopedics. This latter pattern may to some extent be an artifact of selection bias in construction of the sample; for example, it is likely that a more liberal definition of tissue engineering to include a greater part of islet cell transplantation research and a correspondingly more aggressive search for clinicians who have been involved in such work would have resulted in the inclusion of more endocrinologists in the roster. Nevertheless, we would expect surgeons to dominate in any reasonably representative sample of clinician-scientists active in tissue engineering. At present, most of the individuals active in this representative sample of tissue engineers entered the field after completing dissertations in other areas. Because of the interdisciplinary character of tissue engineering, the field will likely continue to have a relatively high proportion of post-doctoral-training entrants. However, over time, the fraction of individuals whose thesis research was explicitly in tissue engineering, and the representation of universities other than MIT with broad platforms of research and training activity in tissue engineering, such as Rice, UCSD and Georgia Tech, should increase as a proportion of any roster of researchers active in TE.

One last characteristic of this cohort is noteworthy here. The great majority of individuals are involved in TE on only a part-time basis. That is, academic tissue engineers typically maintain a number of lines of research, some of which meet any reasonable definition of tissue engineering, some of which straddle the ill-defined boundary that delineates the field, and some of which are clearly situated within other intellectual domains. By this measure, tissue engineering could be viewed more as a tactic than as a discipline – as part of an interdisciplinary assault on unsolved therapeutic challenges that draws opportunistically on an ever-wider range of knowledge and tools from science and engineering.

TISSUE ENGINEERING AND LEADING CENTERS IN THE FIELD

A bibliometric analysis of papers important to the field of tissue engineering revealed a list of more than 350 US institutions active in research. Our interviews with experts in the field as well as extensive secondary research validated this list and also pointed toward several other institutions as being highly influential in the field—not only in terms of their research contributions, but also in their training of personnel currently active in the field.

BOSTON AREA: MIT AND HARVARD

Perhaps the single most prominent geographical and institutional locus of research in tissue engineering has been in the Boston area, centered on the Massachusetts Institute of Technology (MIT) in Cambridge and the Harvard Medical School (HMS) in Boston, although researchers at other Boston-area institutions have been involved as well. MIT was the site of three independent, seminal lines of research on substitutes for human skin. During a period from the mid-1970s through the mid-1980s, the laboratory of cell biologist Howard Green, MD achieved a breakthrough in the cultivation of human keratinocytes, developed it into a method for growing epithelial grafts from a small piece of autologous epidermis, and demonstrated the viability of this method in the treatment of burn victims.

In 1987, this technique became the basis for a startup company called BioSurface Technology that offered cultured epidermal autografts commercially. The company was acquired by Genzyme in 1994 and became Genzyme Tissue Repair, and the service is still offered under the name Epicel. Dr. Green himself has continued his career as a cell biologist at HMS, but has not been prominent in the subsequent development of tissue engineering.

Ioannis Yannas joined the faculty in the mechanical engineering department at MIT in 1966, after completing his PhD at Princeton. From the start, his research focused on the properties of collagen and its role in connective tissues, using approaches from chemistry, physics, and biomechanics. By the 1970s, Yannas was investigating the use of acellular collagen-glycosaminoglycan matrices as wound dressings designed to serve as biodegradable templates for the regeneration of viable skin.

In 1977, Yannas, in collaboration with Massachusetts General Hospital and Shriners Burns Institute surgeon John F. Burke, was awarded a patent for a "multilayer membrane useful as synthetic skin". In 1980, they published the first in a series of papers outlining design considerations for what they called an "artificial skin", and shortly thereafter, they reported the successful use of such an artificial skin in the treatment of extensive burn injury.This research resulted in the development of a commercial product, Integra Dermal Regeneration Template, which is currently licensed to and manufactured by Integra Life Sciences Corporation. Prof. Yannas has continued to maintain an active research programme on the principles and applications of induced tissue regeneration.

Eugene Bell, a long-time faculty member in the biology department at MIT, also pursued in the 1970s a line of research that examined interactions between collagen and skin cells. In 1979, Bell and colleagues published a paper describing the production in vitro of a tissue-like structure, obtained via contraction of a collagen lattice seeded with fibroblasts. In 1981 Bell's group reported the successful grafting of a "living skin equivalent" consisting of fibroblasts cast in collagen lattices and seeded with epidermal cells.

A patent was obtained for this "tissue-equivalent" in 1984, and Bell left MIT in 1986 to found Organogenesis, Inc. and pursue commercial development of the product. He left Organogenesis in 1991 to return briefly to MIT, but left again in 1992 to found Tissue Engineering Inc. Bell has led an active proprietary research programme at TEI Biosciences, focusing on the further refinement of collagen fibre-based scaffolds and the enrichment of these scaffolds with signaling molecules that induce stem cell development and tissue regeneration, but has published little in the peer-reviewed biomedical literature in recent years.

Langer's graduate studies were carried out under the supervision of yet another MIT researcher active during the "prehistory" of TE, Clark Colton of the chemical engineering department. Colton's own PhD, completed in the same department in 1969 under the supervision of Kenneth A. Smith, was entitled Permeability and Transport Studies in Batch and Flow Dialyzers with Applications to Hemodialysis. Colton's research on filtration technologies relevant to artificial organs continued following the completion of his PhD, and he collaborated with William Chick at Harvard Medical School, Pierre Galletti at Brown University and colleagues on the research which led to a 1975 publication in the Transactions of the American Society for Artificial Internal Organs on a "hybrid artificial pancreas" for rats consisting of beta cells cultured on synthetic semipermeable hollow fibers.

Colton was the only senior MIT researcher represented with a paper at the Granlibakken workshop in 1988. Although contemporaneous with this early work on artificial organ technologies in Colton's lab, the research that formed the basis for Langer's 1974 ScD in chemical engineering was part of a separate line of work being pursued by Colton in the area of enzyme engineering. Langer proceeded to a postdoctoral fellowship in the laboratory of Dr. Judah Folkman at Harvard Medical School and Children's Hospital in Boston.

Folkman's pathbreaking work on implantable drug delivery systems and on angiogenesis in cancer addressed a number of fundamental issues that were to be of broader significance for tissue engineering, including identification and analysis of the factors that govern the growth and differentiation of blood vessels, and the design of delivery systems to enable the sustained release of bioactive proteins and other molecules in a desired location in order to influence tissue growth.

Langer's entry into tissue engineering was catalyzed by a collaboration with Jay Vacanti. Vacanti earned his MD at the University of Nebraska, and came to Boston in 1974 to begin his postgraduate training in surgery. He completed a residency in general surgery at the Massachusetts General Hospital during 1974- 1981, followed by a fellowship in pediatric surgery at Children's Hospital from 1981-1983. During the period of his residency at MGH, he spent two years – from 1977 to 1979 – in the laboratory of Dr. Folkman, where his collaboration with Robert Langer began. Their first joint publication, written in collaboration with Folkman and two other colleagues, reported on experiments in the inhibition of tumor growth in animal models by regional infusion of a cartilage-derived angiogenesis inhibitor.

Returning to MIT following his postdoctoral work with Folkman, Robert Langer built a highly productive research and training programme focused primarily on polymer-based drug delivery systems and their applications. Vacanti, following his pediatric surgery fellowship, received two further years of clinical training in the renowned transplantation programme at the University of Pittsburgh, then returned to Boston in 1985 to join the staff at Children's Hospital and launch his own clinical and research programme. Vacanti's clinical experience in transplantation in Pittsburgh, and then in launching a pediatric liver transplantation programme in Boston, confronted him with the intractable problem of organ supply. Following his return to Children's Hospital in Boston in 1985, he began to explore with Langer the problem of how to extend to three dimensions the early success of Eugene Bell in growing flat sheets of tissue from skin cells. From this emerged the concept of using a porous, resorbable, three-dimensional polymer scaffold as a template for cell growth, and the 1988 paper the Journal of Pediatric Surgery reporting its experimental application.

This technique was to be enormously influential in shaping the work of other investigators who entered the field – indeed, it could be said that it provided the intellectual "scaffolding" for the work of an entire cadre of investigators who shared the long-term vision of creating "replacement" solid organs for therapeutic purposes. Although Langer himself is not primarily a tissue engineer, TE projects have remained an active line of research in his laboratory.

He has mentored an extraordinary number of doctoral and postdoctoral trainees, some of whom focused their work specifically on tissue engineering applications. Many have gone on to establish independent academic careers in which tissue engineering research has played an important part.

INSTITUTIONAL SUPPORT OF TISSUE ENGINEERING

In addition to the individual researchers in academia and industry, key players in a field can include the federal agencies that provide financial support, professional societies and other institutions designed to promote and advance a field's mission. Professional societies may promote educational programs, research applications, and the development of professional standards in an overall effort to advance the dissemination of knowledge in a particular field or research area. As networking facilitators, conferences, meetings, and other symposia supported by societies may be instrumental in bringing together researchers who work on similar topics but may not otherwise collaborate.

Several groups have emerged in tissue engineering's development.The National Science Foundation is one of many organizations – including the U.S. Federal government and foreign governments, industry and non-profit foundations – that have funded tissue engineering research in the United States and around the world. The WTEC report estimates that nearly $3.5 billion dollars have been spent on tissue engineering in the last decade. Of this, less than 10 per cent has come from the U.S. Federal government. According to their definition, the National Science Foundation, primarily through its Directorate for Engineering, provided less than 1 per cent of worldwide support for tissue engineering, but nearly 7 per cent of U.S. Federal government support. The accompanying examine U.S. Federal government funding of tissue engineering in greater detail, using a more restrictive definition. They provide a partial glimpse into Federal support of TE using data from RaDiUS, a database that tracks the research and development activities and resources of the Federal government.

Of all Federal agencies supporting biomedical science and engineering, NSF is second only to NIH in providing financial support to TE, and has increased both its level and share of Federal funding in the past few years. Before turning to NSF's role in supporting tissue engineering, we provide brief descriptions of major efforts of other Federal agencies that have been active in tissue engineering in different ways.

National Institute for Standards and Technology (NIST)

NIST's Advanced Technology Programme (ATP) began in 1988 and was designed to focus on commercial applications of high-risk cutting edge basic scientific research, such as biology, a field NIST did not yet participate in. The first ATP competition was held in 1990-91 and the first TE award was made in the 2nd round of competition to Acerm Biosciences, a company growing stems cells from bone marrow. In the years subsequent, NIST added projects at the rate of 1-2 per year on topics such as building better scaffolding, bioreactors, and xenotransplantation.

Serious efforts in tissue engineering, however, did not occur until the mid-late 1990s. In 1993-94, a two day conference was held on cutting-edge biotechnology and involved individuals from several agencies including Francis Collins of NIST, Fred Heineken of NSF, and Kiki Hellman of FDA. The group identified 'hot topics' in a hierarchy—tissue engineering being one of those near the top of the list. For each, a set of key issues was identified and listed: in what areas is there enough basic science research and where is there a black hole? Other technologies identified at this meeting included DNA diagnostics, TE, vaccines, gene therapy, and biosensors. Approaches for the development of specific industries were discussed. Stan Abramowitz, who headed the early NIST/ATP efforts, was heavily involved in this. Staff attended conferences to "find the right people" to get advice, and to head these initiatives. As a follow-up to these efforts, a series of white papers were solicited in 1995 in preparation for a focused competition in TE. The first focused programme was initiated in 1997 when 56 proposals were submitted and 12 awarded (for a total of $12 million that year).

Focused competitions in tissue engineering ended for ATP in 1998, however. The ATP Programme has been instrumental in funding high-risk commercial applications of ideas that venture capitalists are unlikely to support—typically these are multidisciplinary projects which require academic collaboration. Dr. Rosemarie Hunziker, a former Programme Officer at ATP, believes that NIST/ATP support of TE gave it credibility, an important action, especially in light of its heavy private sector orientation.

National Aeronautics and Space Administration (NASA)

Beginning in the late 1970's, NASA's efforts in tissue engineering have focused on development of techniques for three-dimensional cell and tissue culture. The agency's interest in tissue engineering comes from a need to understand how cells and tissue behave in space. Laboratory cultures which mimic live human tissue can be used as models to conduct space research and research on deadly diseases, such as cancer—eliminating the need for human test subjects. In the late 1980's, the work of Milbourne and Wolf of the Johnson Space Centre culminated in the development of the rotating bioreactor, a horizontal rotating wall vessel with a centre oxygen membrane, which promoted cell and tissue growth under a variety of conditions. A patent on the device was filed in 1988 and issued in 1991.

While initially referred to under the agency's biotechnology heading, NASA officially recognized these efforts as "tissue engineering" when the term appeared for the first time in a 1994 programme update. Since then, NASA has continued to fund selective research on bioreactors and related topics and has provided a steady stream of funding to individuals in this area. Much of the work done by Lisa Freed and Gordana Vunjak-Novakovic of MIT has been supported by NASA's Cellular Biotechnology Area.

Food and Drug Administration (FDA)

The FDA has taken an interest in tissue engineering since the early 1990's. In 1990, the Centre for Devices and Radiological Health (CDRH) in conjunction with the State of Maryland held a conference to examine biological applications with device use. Among the topics raised was tissue engineering. Dr. Kiki Hellman, head of the CDRH, saw potential in the field and became interested in exploring it further.

Two years later, a workshop was held to examine promising new technologies—tissue engineering being one of them. From an agency perspective, the FDA was proactively involved in the technology's development. According to Dr. Hellman, the FDA made a concerted effort to deal with in-house needs (staff training, education) so that they could anticipate the technology and be prepared to deal with the any technical or scientific issues that would arise.

Many of our interviewees also state that the FDA has been instrumental in allowing products to move to market, encouraging the development of standards to measure devices (in conjunction with the ASTM Committee F04 Medical and Surgical Materials and Devices). In all, many experts in the field laud their efforts, stating that the FDA is not a force standing in the way of progress.

The Tissue Engineering Society International (TESI)

In 1995, Charles Vacanti started The Tissue Engineering Society. In its early days, the society served to foster communication between individuals in the tissue engineering world through sponsorship of meetings and conferences. In 1995, the Society published the first issue of Tissue Engineering, the first dedicated journal in the field. Despite the placement of several prominent tissue engineers on its editorial board, the journal lacked popularity. Many researchers continue to preferentially publish in more traditional disciplinary journals or general peer-reviewed publication like Science or Nature. More recently, the society has grown to include a more international focus and serves as a networking organization to foster dialogue between international researchers in the field.

The Pittsburgh Tissue Engineering Initiative

A unique and somewhat unusual contributor to the field of tissue engineering has been the Pittsburgh Tissue Engineering Initiative (PTEI). The organization was founded by Peter Johnson a former surgeon at the University of Pittsburgh. After becoming chairman of the department of plastic surgery, Johnson was in a unique position to form a network of scientists in the Pittsburgh area. Based largely upon his own interest in tissue engineering and its potential, Johnson pushed to develop a joint technology transfer policy between the major Pittsburgh area universities, and PTEI was born. In its beginnings, PTEI served mostly as a networking mechanism—to raise funds to provide support to TE research in the surrounding area. The initiative had four key components:

- Technology development grants,
- Summer research internships to increase student base in TE,
- Biotech exposure (with a minority focus),
- A guest speaker programme (a coordinated effort to bring big names in the field to Pittsburgh).

PTEI raised money and support and led the 1st TE biotech company in the area. Currently, there are at eight firms in the area. PTEI has encouraged an amalgamation and growth of the critical mass necessary to sustain development in TE for the Pittsburgh community. As a result, Pittsburgh is rapidly becoming a centre for informatics—a new branch in the field of tissue engineering. PTEI also serves as a model for other cities to advance research and development in TE. Toronto is expanding their TE workforce utilizing the PTEI model.

The Whitaker Institute for Biomedical Engineering

The mission of The Whitaker Foundation is to promote better human health through advancements in medicine. This is accomplished through a series of competitive grant programs that support research and education in biomedical engineering at academic institutions in the United States and Canada. Whitaker has been instrumental in funding bioengineering programs around the country. Most notable for TE, they, in 1999, established the Whitaker Institute for Biomedical Engineering (WIBE) at UCSD under the leadership of Y.C. Fung. Much of the fundamental work supported under this institute will contribute to advancements in the field of TE. However, the foundation has been reluctant to classify TE as an independent entity, but rather a sub-field of bioengineering.

Multi-Agency Tissue Engineering Sciences (MATES) Working Group

More recently, many agencies of the federal government have become interested in supporting the mission of tissue engineering. In an effort to coordinate this support, representatives from each of several federal agencies came together in 2000, under the leadership of Drs. Fred Heineken (NSF) and Kiki Hellman (FDA) to form the Multi-Agency Tissue Engineering Science (MATES) Working Group. MATES has three major goals:

- To facilitate communication (and prevent redundancy of funding) across departments/agencies by regular information exchanges and a common web site,
- Enhance cooperation through co-sponsorship of scientific meetings and workshops, and facilitate the development of standards,
- To monitor technology by undertaking cooperative assessments of the status of the field.

MATES has developed a web-site which they hope will become "onestop-shopping" for those seeking information on the field (such as information about federal funding, scientific meetings, regulatory guidance and standards development).

Participating agencies in MATES include the Department of Commerce (NIST), the Department of Energy (DOE), the Department of Defence (DARPA), the Department of Health and Human Services (FDA, NIH), the National Aeronautics and Space Administration (NASA), and the National Science Foundation (NSF). The panel, co-chaired by Drs. Frederick Heineken of NSF and Kiki Hellman of the FDA also focuses on some regulatory, legal, and ethical issues for TE. A particular area of interest has been a global regulatory initiative for TE—an effort that will require the development of standards for testing tissue engineered products.

A major contribution of the MATES group is the recent World Technology Evaluation Centre (WTEC) study on tissue engineering, which attempts to summarize the technical contributions of current work in TE and to estimate the funding for TE from each of the federal agencies. The study has become a focal point for activities of the working group, under the auspices of the Subcommittee on Biotechnology, Committee on Science of the President's National Science and Technology Council (NSTC). The results of the WTEC study were used by MATES to plan a joint interagency programme announcement in tissue engineering issued under the new National Institute for Biomedical Imaging and Bioengineering, which solicits research aimed at addressing specific gaps in tissue engineering.

The Role of the National Science Foundation (NSF)

The National Science Foundation, particularly through its Directorate for Engineering, appears to have played an important role in the emergence of tissue engineering as a recognized field of activity, and in shaping the character and in determining the direction of the field.

Financial Support

From an examination of Fastlane Award data as an illustration of NSF support for tissue engineering, it is evident that the majority of NSF support goes toward individual investigator awards and Centers research. NSF is also a strong supporter of workshops, conferences and other meetings, which suggest a prominent role in support of networking activities among individuals active in the field. Analysis of NSF funding by division shows that the vast majority of the agency's support of tissue engineering—more than 80 per cent—emanates from the Bioengineering and Environmental Systems (BES) and Engineering, Education and Centers (EEC) Divisions—both of which reside under NSF's Directorate for Engineering.

Researcher Support

While the interviews did not shed much light on the role of NSF as a funder, bibliometric analysis revealed that NSF has had a significant role in supporting lead researchers in the field. The relative role of NSF support of TE over the years. NSF support has been acknowledged on 12 per cent of the papers revealed by the bibliometric analysis. The influence of NSF funding on top researchers in the field, however, has been disproportionately large. Three of the most prolific researchers – Vacanti, Langer and Mooney - acknowledge NSF funding on 19 per cent-37 per cent of their papers, including on the

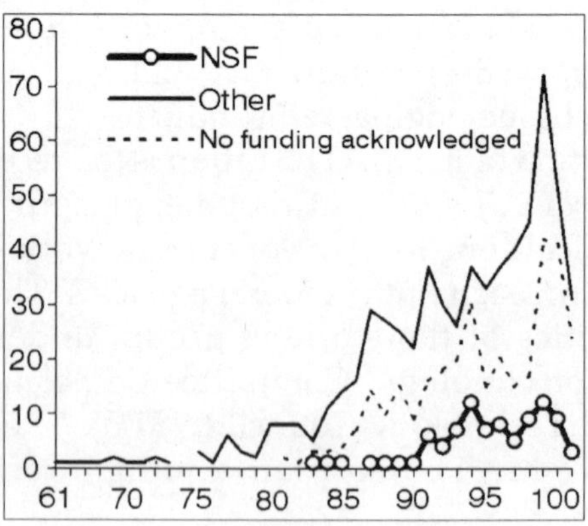

1993 Science paper (this percentage may, in fact, be even higher, as not all papers list acknowledgement of research funding and, further, some researchers fail to mention all of the funding sources that should be cited in acknowledgements). Given the wide range of tissue engineering activities, it is not surprising that NSF support of researchers varies considerably.

Notable researchers such as Linda Griffith, Jeffrey Hubbell, P.M. Kaufmann, and Mark Saltzman also acknowledge NSF on a third or more of their papers, while NSF appears to have had no role in supporting other prominent researchers such as Arnold Caplan, Lisa Freed, V.M. Goldberg or Gordana Vunjak-Novakovic. NSF funding, however, is not evenly distributed across all aspects of tissue engineering. NSF support is clearly targeted toward the application of engineering principles, especially with the aim of advancing engineering knowledge. The bibliometric study to identify the sub-fields of tissue engineering where NSF-funded researchers have been most active and demonstrates that the agency has held true to its stated mission and goals. Almost 90 per cent of NSF-supported papers are in fields related to engineering or fields that promote the application of engineering principles to solve problems in a variety of realms.

Institutional Development

NSF has played a small but significant role in developing institutions and networks in the field of Tissue Engineering.

Conferences and Workshops

NSF has been an early sponsor of workshops and conferences related to TE. The term "tissue engineering" itself was proposed at a 1987 panel meeting convened to consider future directions for the Engineering Directorates's Bioengineering and Research to Aid the Handicapped Programme. Strong interest in this concept within the Engineering Directorate led to a special panel meeting on tissue engineering, again at NSF, in the fall of 1987, and then to the Granlibakken workshop of 1988, now considered to be the first formal scientific meeting of the emerging field of tissue engineering. In all, about one per cent of NSF's total TE support in the period 1987-2001 was devoted to networking and conference support, and many of the individual investigator awards in TE were funds to support presentations of results from research conducted with the NSF support and attendance at non-NSF conferences and meetings.

MATES

As a founder and critical participant in the MATES Working Group, the Engineering Directorate, through the leadership of Frederick Heineken, Kiki Hellman, and others, has made a pronounced effort in recent years to engage participation from a range of disciplines. The MATES group aims to enhance collaboration and cooperation amongst several federal agencies so that gaps in tissue engineering research can be addressed. An important example is an outgrowth of the MATES initiative, a study published through WTEC, which summarizes the global status of the field and notes where progress is needed.

NIBIB

Aside from its participation in MATES, NSF continues to collaborate with other Federal agencies and programs on TE related research and education. In particular, NSF has ongoing collaboration with the newest of the NIH Institutes, the National Institute for Biomedical Imaging and Bioengineering (NIBIB), established in 2000. The first "official" NIBIB interagency activity was a joint NIH/NSF Workshop on Bioengineering and Bioinformatics Education and Training.

Since then, the two agencies have established the Bioengineering and Bioinformatics Summer Institutes (BBSI) Programme to provide students majoring in the biological sciences, computer sciences, engineering, mathematics, and physical sciences with interdisciplinary bioengineering or bioinformatics research and education experiences137.

Another important outgrowth of the NIBIB is a recent solicitation (dated December 30, 2002), likely influenced by the work of MATES and the WTEC report, seeking proposals to address key research challenges in TE138. The initiative is expected to draw biologists, clinicians, and engineers to work together to address issues such as cell sourcing, cell identification and characterization, engineering design, and other enabling technologies necessary to move the field forward. A total of $8 million dollars has been earmarked for the initiative.

Centre Support

NSF's Directorate for Engineering has provided significant support toward institutional development for TE through its funding of Centers. These include the Georgia Tech/Emory ERC, which was awarded in 1998, and the University of Washington Engineered Biomaterials (UWEB) Centre. Aside from NSF, the bulk of institutional support in TE has been provided by the Whitaker Foundation, through its role in building the field of biomedical engineering, and especially in providing institutional development funds for the creation and expansion of bioengineering departments– not only at the academic centers highlighted, but at many other institutions across the nation.

These departments are increasingly taking over from departments of chemical and mechanical engineering as the focus of academic research activity in tissue engineering, and the drive to create and expand TE focus areas in many of these emerging departments has created opportunities for young investigators completing their training at the institutions that have led the way in TE research. Not evident in these funding analyses, but worthy of note, is that NSF has been an important source of support over the years for the work of Douglas Lauffenburger, a highly productive researcher on the fringes of tissue engineering who has trained many younger investigators involved in the field. Since 1999, Lauffenburger has served as director of the MIT Biotechnology Process Engineering Centre (BPEC), an NSF-funded ERC, in its second incarnation in 1995. BPEC has developed new research thrusts that have moved the focus of its activity toward the boundary between tissue engineering and several adjacent fields, with projects led by several MIT investigators prominent in tissue engineering.

While available evidence suggests that no single organization engendered monumental changes in the field of tissue engineering, NSF's Directorate for Engineering did create an environment that encouraged the crystallization of an emerging concept, both with its innovative Engineering Research Centers (ERC) Programme and its ongoing internal efforts to define productive future directions for its bioengineering programme. NSF hosted the first meetings at which the concept of tissue engineering as a focus of research was defined and strategies for its future growth and potential discussed.

While the Directorate for Engineering has no doubt provided funding support to key researchers in the field, it is more difficult to attribute specific research contributions and breakthroughs to the Directorate. In fact, in general, the broad and interdisciplinary nature of the field make it difficult to classify any number of research contributions as seminal. Efforts to "engineer" tissues to restore structure or function predate the events of 1987-88 by many years. However, if there is a research development that marked the emergence of a recognizably distinct field on this foundation of many related, pre-existing lines of research, it appears that the consensus in the research community and through our own best judgment points to the January 1988 publication by Langer, Vacanti and colleagues describing the strategy of using resorbable artificial polymer matrices seeded with cells as a vehicle for cell transplantation. This early phase of the Langer/Vacanti collaboration, catalyzed by the shared experience of working under Folkman, brought together at least four of the conceptual elements that are central to tissue engineering today: the use of polymeric materials in biological applications, the use of live cells to restore lost physiologic function, the vision of controlling tissue morphogenesis through the use of signaling molecules, and the powerful motivation of the shortage of transplantable organs.

Further, the seminal development in awareness of tissue engineering in the scientific community as a distinct domain of research is also attributable to Langer and Vacanti. The October 1987 Panel Meeting at NSF and the subsequent Granlibakken workshop in 1988 and Keystone workshops in 1990 and 1992, while significant, appear to have "spoken" to a much narrower audience; among our group of interviewees, few who were not directly involved in one or the other of these events report any awareness of their direct impact, although, again, the field definition drafted at the October meeting is acknowledged. NSF's Directorate for Engineering also appears to have played a large role in incorporating the biomechanics community into the emerging field. The events of 1987-88 engaged pioneers of an engineering approach to orthopedics and hemodynamics, such as Richard Skalak, Van Mow, and Robert Nerem, in the development of a shared concept for the field just as the emergence of the scaffolds-and-cell-seeding approach, occurring in parallel with and independently of the Foundation's activities, was providing a renewed and powerful impetus for efforts to construct cell-based therapeutic solutions.

One might also note the delayed efforts of the Foundation to engage biologists in the endeavor. All of the participants in the Foundation's 1987 discussions were well-aware of the importance of fundamental research in biology as part of the mix required to make tissue engineering a success. As an agency concerned with the full range of fundamental research and not directly with clinical applications, NSF might have been the natural locus for an effort to mobilize basic biological scientists as well. However, as an NSF initiative, tissue engineering arose from the engineering directorate, and judged by data available from the NSF Fast Lane grants database, remained largely contained within the engineering directorate.

As noted earlier, tissue engineering continues to be characterized by a tension between ad hoc and more fundamental or systematic approaches, with many observers concerned that the balance remains too far toward the ad hoc. On this deeper level, the involvement of engineers should be an important contribution to making TE truly an engineering discipline, characterized by rational design based on integration of fundamental principles. In practice, however, this theoretical rationalization has not made much progress. The challenges addressed by tissue engineering are hard problems. This perception underlines a dilemma faced by the Foundation as a whole. In the future, NSF will be challenged to define and fulfill a role that allows it to continue to have a noticeable impact on the field.

3

Technology of Tissue Culture Propagation

PRIMING TISSUE CULTURED PROPAGULES

Priming of in vitro propagules refers to the manipulation of the growing environment, prior to and upon transplanting, and is an integral part of tissue culture propagation. Tissue-cultured propagules are produced under controlled environment. Such plantlets have small juvenile leaves with reduced photosynthetic capacity, and malfunctioning stomata. Priming for rooting, shooting, and improved photosynthesis can be achieved with growth regulators and simple adjustment to the growing conditions that affect the post-transplanting performance of the propagules.

Vented closures with microbial filters have been used to facilitate gas exchange to reduce ethylene build-up that stunts plant growth, reduces leaf-size, and causes leaf drop in tissue culture containers. Plantlets produced under photoautotrophic culture systems on media with or without sucrose but CO_2-enrichment, increased light intensity, good gas exchange and reduced humidity are more vigorous, have larger root- systems, and are less susceptible to microbial contamination. Plants adapted gradually to the ex-vitro environment have improved survival upon transfer to soil. Plants in their natural environment are colonized with many bacteria, fungi, and mycorrhizae. In-vitro or ex-vitro biopriming of micropropagated plants with such organisms can improve plant performance under stress environments, and consequently enhances yield.

The adaptation of plants to the environment - temperature, osmotic stress, and pathogens, is triggered by modification of expression of specific genes and metabolic pathways. Certain chemicals, environmental factors, and microorganisms pre-sensitise the cellular metabolism of plants. Upon exposure to stress, the pre-sensitised or primed plants adapt better and faster than the non-primed plants. The production of high quality and vigorous plants through in vitro culture requires the enhancement of post-transplanting ability for water management, efficiency of photosynthesis and resistance to diseases. Priming of in vitro propagules, based on manipulation of the growing environment (chemical, physical and biological) prior to and upon transplanting, is an integral part of tissue culture propagation. The ability of the propagules to withstand transplanting stress very often determines the success or failure of tissue culture operations.

Tissue-cultured propagules are produced in controlled environments under pre-set temperature and low-light intensity, and in small vessels with poor air exchange. This leads to high humidity and accumulation of ethylene. Also, the nutrients in the growing media, sucrose, mineral salts, vitamins and growth regulators, are far in excess of those in the soil. Such an environment causes developmental distortions, and represses or modulates several metabolic pathways. As a result, the plantlets have small juvenile leaves with reduced photosynthetic capacity, and malfunctioning stomata. The roots of such plants lack hair or have few, and the leaves have poor cuticle development and low wax deposit.

Priming for Rooting

The ability to manage water, photosynthesis and response to stress during and after in vitro culture determines the final performance of the propagules. The formation of a functional root system is an essential step to manage water under the post vitro conditions. The in vitro roots remain functional and continue to grow during ex-vitro acclimatisation. Easy to root species, (e.g. potato) usually regenerate roots upon subculture on hormone-free medium.

However, root architecture and biomass partitioning between roots and shoots can be easily manipulated by simple adjustment to the growing conditions that affect the post-transplanting performance of the propagules. For example, increasing day/night temperature from 20/15 to 33/25°C either slightly stimulates or has no effect on the shoot growth in clones of potato, but dramatically enhances root growth and root branching.

Several chemicals also influence the development of the root system. Abscisic acid is known to affect initiation of new lateral roots and overall changes to the root architecture Jasmonic acid (JA) has a similar effect. Low JA concentrations (0.5- 1.0 μ M) stimulate both root and shoot growth. Jasmonates induce stress resistance in plants, and may be useful for priming of in vitro plants during the last subculture.

In potato, nodal explants from JA primed stock plants tuberize earlier and more uniformly, and produce more microtubers than the non-primed controls. Salicylic acid (SA) is a major signalling compound and induces abiotic and biotic stress resistance in plants. It also induces adventitious roots in cuttings and stomata closure, and inhibits ethylene biosynthesis in detached leaves; hence it may also be useful for priming tissue-cultured propagules. Many woody and herbaceous perennials require auxins for rooting. Auxins are involved in root initiation and adventitious root formation. However, excessive auxins after root induction may inhibit root regeneration and elongation, and impair formation of a good root system.

Therefore, a careful selection of the auxin type and concentration is recommended. IAA and IBA are usually used for easier-to-root plants and NAA for more recalcitrant woody plants. The efficacy of different auxins also depends on the explant type, and exposure to light. For example, a short duration (3-7 days) of dark incubation enhanced rooting in apple, and papaya. Addition of low concentration of the growth retardant daminozide to hormone-free medium enhanced post-transplanting acclimatisation of potato plants.

Other supplements like activated charcoal, seaweed concentrate, pyroligneus acid, and dilution of salts in the basic MS medium can also aid root initiation and growth. In vitro cultured papaya shoots in aerated vermiculite gave better rooting with higher post-transplanting survival than on non-aerated vermiculite and non-aerated agar. In some difficult-to-root species, e.g. apple and Saskatoon berries, in vitro rejuvenation via several subcultures is required before successful rooting of microcuttings. Individual cultivars may require adjustment to culture conditions and different number of subcultures. For example, over 90 per cent rooting was achieved with apple microshoots, cv. 'Jonathan' by the ninth subculture, but a more recalcitrant variety 'Red Delicious' required over 30 subcultures for successful rooting. In Saskatoon berries, where rooting by traditional stem cuttings is very difficult, tissue culture rejuvenation and subsequent transplanting to the greenhouse can generate highly vigorous plants. Shoot cuttings taken from such planted hedges readily root within 2-3 weeks after a simple treatment with IBA.

Fig. Five-week-old Potato Plants cv. 'Kennebec' Grown on 10 ml Hormone-Free MS medium (pH 5.7, 3 per cent Sucrose, Vitamins and 0.6 per cent agar), Supplemented with various Concentrations of JA at 22/18ºC day/night Temperature, 16 h Photoperiod, 150 μ E-2s^{-1} Fluorescent Light

Increasing Photosynthetic Competence

Tissue culture containers provide sterile conditions and high moisture, thereby preventing contamination and desiccation of the explant, but restrict aeration and escape of metabolically generated ethylene and water vapors. Such conditions slow down growth and induce undesirable morphological changes. The ethylene build-up stunts plant growth and reduces leaf-size, and causes leaf drop. Such plants desiccate quickly upon transfer to the external environment. Vented closures, including those for baby-food-jars with built-in different sizes of microbial filters, have been used to facilitate gas exchange between culture vessels and the ambient environment.

High relative humidity in culture vessel reduces deposition of epicuticular waxes on leaves. It impairs the functioning of stomata and alters leaf mesophyll by reducing the number of palisade cells and by forming spongy parenchyma with large air spaces. Other effects include low nutrient uptake due to decreased transpiration, decreased lignification, and reduced or abnormal development of the vascular system. Although the cuticular wax development and stomata functioning can be induced by gradual adjustment to lower humidity, morphological changes cannot be easily reversed. The altered leaf morphology reduces photosynthesis and enhances respiration. Cooling the bottom of culture vessels by 2-3°C below the ambient air temperature in the culture room or providing high light intensity can reduce high humidity. Gradual opening of vessels for a few days prior to transplanting can also adjust high humidity.

Many micropropagated plants have C3 photosynthetic pathway; hence, are greatly affected in their ability to assimilate carbon. The balance between photosynthesis and photorespiration controls their productivity. Under the traditional methods of micropropagation, cultures are usually mixotrophic (use carbon source mostly from sucrose in the medium as well as photosynthesis). The carbohydrates in the media inhibit carbon assimilation via photosynthesis by reducing rubisco (ribulose-1-5-bisphosphate carboxylase/oxygenase) activity. On the other hand, several studies indicate a stimulatory effect of sugar on biomass accumulation under elevated CO_2. In the photoautotrophic culture systems the cultures are grown under CO_2-enrichment (1000-1500 μ mol mol^{-1}), increased light intensity (at least 150 μ mol m^{-2}, s^{-1}), good gas exchange and reduced humidity, and on media with or without or sucrose. Plantlets produced under such conditions are more vigorous, have larger root- systems, and are less susceptible to microbial contamination. They are able to photosynthesize, and do not require acclimatization during post-vitro establishment.

PRIMING FOR TEMPERATURE TOLERANCE

In vitro grown plantlets can be acclimatized to low temperatures for cold tolerance under field conditions. Alfalfa and red clover seedlings withstood -

16 and -10° C, respectively, upon cold hardening at 2-5° C, 80ìE m^{-2}, s^{-1} light and 8h photoperiod, for 3-5 weeks. Such hardening has been successfully utilized in the selection of cold tolerant alfalfa and red clover genotypes. Elevation of sucrose in the culture medium or application of abscisic acid, or both, may further improve plantlet hardiness in some species.

ACCLIMATIZATION TO SOIL TRANSFER

Good quality propagules with well developed roots and leaves are easy to acclimatize to the external environment. Any successful acclimatization protocol must ensure that the plants maintain active growth during the entire weaning period. The effect of in vitro growth conditions on post-vitro acclimatization has been demonstrated in grapevine and chile ancho pepper, Capsicum annuum. The most common approach to improve plant survival upon transfer to soil is their gradual adaptation to the ex-vitro environment. Under such conditions, plants rapidly convert from heterotrophic or photomixotrophic to autotrophic growth, form fully functional root systems, and their stomatal and cuticular transpiration is reduced.

Such gradual adaptation is often carried out in greenhouses by decreasing relative humidity using fog or mist chambers and increasing light intensity and shading techniques. Light intensity needs to be adjusted to the plant requirements. Rooting of in vitro produced shoots in vivo is recommended for some species to promote development of roots already adapted to external conditions, to prevent potential damage to fragile in vitro roots, or to avoid post-rooting dormancy. Other treatments of in vitro plantlets upon transplanting to the outside environment include high phosphate fertilizer to enhance their vigour, fungicides to prevent damping-off and anti-transpirants to prevent desiccation; however, they may cause phytotoxicity.

A simple, low cost method for ex vitro weaning and liner production of woody plant species and Hosta has been developed. The shoots with root-initials from the rooting medium are transplanted in 'Jiffy' plugs. Plants are acclimatized in plastic trays (25x50x5cm) with transparent cover. Each tray holds 300 plugs, each 18mm diameter. The plugs are saturated in a 2.5-3L solution of fertilizer (1g/L) 10-52-10 NPK; the excess solution is discarded before transplanting. The plants are watered weekly (approx. 1L water/week) and kept in the growth room benches for 3-weeks after transplanting at 22-24/20ºC day/night, 14h photoperiod, under 120 μ $Em^{-2}s^{-1}$ fluorescent light. The current cost is US$73 for 5400 plugs, $58 for 100 trays, and $60 for 100 plastic covers; the total cost per plug being US$0.018. The rooting success in woody plants, e.g. Swedish columnar aspen, Saskatoon berry, chokecherry and pincherry cultivars, sour cherry, low bush blueberries and others was more than 95 per cent, and 100 per cent with hosta varieties. In seed potato production, 2 to 4-week old tissue cultured plantlets are usually hardened off for 2-4 weeks in a greenhouse, and transplanted in the field. Such plantlets are still juvenile and not well adapted to stress. To prevent desiccation, insect and bird damage, and spread of viruses by aphid, many growers cover plants with floating covers and install an overhead irrigation system. Seed potatoes produced under floating covers are consistently better in quality than those from the unprotected planting.

Fig. Plant acclimatization in plastic trays with transparent cover. Left. Trays in a growth room bench next to culture vessels. Right. Aspen (Populus tremula var. erecta) liners in Jiffy plugs, three weeks after transplanting

In vitro produced-propagules of woody species, e.g. Saskatoon berry (Amelanchier alnifolia Nutt.), Chokecherry (Prunus virginiana L.), Pincherry (P. pennsylvanica L.), Swedish columnar aspen (Populus tremula var. erecta), European weeping birch (Betula pendula var. 'Gracilis' become dormant after rooting. In vitro rooted Saskatoon berry is particularly prone to dormancy as stress factors readily induce terminal bud dormancy. When the shoots stop growth, leaf shedding occurs in the culture or soon after transplanting, and the propagules desiccate. The liners produced from such plants are less vigorous and difficult to establish in the field.

In Chokecherry (P. virginiana L.), dormancy symptoms of shoots with reddish leaves and cessation of terminal-bud growth are induced upon depletion of nutrients in the medium. Post-rooting dormancy has also been reported in rhododendrons and low bush blueberries. Adding low amounts of gibberellic acid (GA) to the medium prevents dormancy, although shoots treated with GA are difficult to root. In Saskatoon berry, ex-vitro rooting under non-sterile conditions is helpful, although does not eliminate the problem. Spraying leaves of ex vitro rooted-shoots with BAP (400ppm), followed by GA (100-PPM) the next day prevents dormancy.

BIOPRIMING SOMATIC EMBRYO-DERIVED PLANTS

Somatic embryos are produced as adventitious structures directly on explants of zygotic embryos, from callus and suspension cultures. The stress inducing molecules in plants, ABA and JA have been successfully used to facilitate embryo maturation and priming for resistance to post transplanting stress. ABA and osmotic shock with 6 per cent sucrose is a routine step to induce desiccation tolerance in alfalfa synthetic seeds. Osmotic stress induction with 3 per cent sorbitol was effective in embryo maturation and conversion in soybean cultures and ABA treatment combined with partial desiccation of encapsulated somatic embryos in sugarcane. Several sugar alcohols and polyethylene glycol have been tested with different degrees of success for priming SE propagules of various plants. In conifers, post ABA embryo maturation has been significantly improved by benzyl adenine (BA). The lowering of sucrose concentration from 58 to 29mM in the proliferation medium improved the formation of cotyledonary SE on the maturation medium. Interaction between plant growth regulators and various carbohydrates and organic nitrogen compounds in embryo development and maturation has been reported.

BIOPRIMING

Plants in their natural environment are colonized both by external and internal microorganisms. Some microorganisms, particularly beneficial bacteria and fungi, can improve plant performance under stress environments, and consequently enhance yield. Plants infected by microorganisms develop systemic resistance (systemic acquired resistance, SAR, or induced systemic resistance, ISR), and/or benefit from their antagonistic abilities towards pathogens (cross protection).

Although, the inoculation of seeds with beneficial microorganisms has been practiced for more than 50 years, the inoculation of tissue culture propagules to enhance plant performance is relatively new. Plant tissue culture is based on axenic (contaminant-free) culture systems. Hence, microorganisms, including endophytes, are treated as problemcausing contaminants, and various procedures have been developed to eliminate them.

Only recently, have microbial inoculants, primarily bacterial and mycorrhizal, been evaluated as propagule priming agents both as in vitro co-cultures and on transplanting. Standard microbiological techniques allow culturing only a few per cent of the naturally occurring microorganisms. Moreover, the soil and plant associated bacteria can switch between culturable and unculturable stages. The development of new culture methods that allow the establishment of stable associations between plants and beneficial organisms, both in vitro and ex vitro, under different environments is a major challenge of future research.

In Vitro Bacterization

Rhizosphere bacteria interact with plants and other inhabiting organisms by producing a wide array of chemicals. Many of these chemicals exhibit plant regulatory activity. There is increasing evidence that the bacteria associated with plants (endophytic and epiphytic) induce host resistance to environmental stresses. This is commonly linked to the production of salicilate or jasmonate, or both. These two types of signalling chemicals are associated with resistance to biotic stress. Recent molecular evidence in Arabidopsis thaliana inoculated with the plant growth promoting rhizobacterium,

Paenibacillus polymyxa, indicates however, that the abiotic and biotic stress-resistance are linked. An effective plant beneficial bacterium, Pseudomonas spp. strain PsJN, was used to enhance tolerance to transplanting stress in potato. The bacterium was originally isolated as a contaminant from Glomus vesiculiferum infected onion roots. A key element in its isolation was that the bacterium (Pseudomonas sp. strain PsJN) did not grow on a standard potato culture medium in the absence of plantlets. Pre-selection of plant beneficial bacteria on plant tissue culture media is thus recommended as a critical step in the development of an in vitro co-culture system, so the bacteria do not overgrow the culture.

Dipping potato nodal cuttings in a suspension of a beneficial bacterium can dramatically affect in vitro growth of plantlets. So far, our isolate, Pseudomonas sp. strain PsJN, has been the most effective plant growth promoting bacterium under in vitro conditions. It forms endophytic and epiphytic populations when co-cultured either with potato, tomato and grapevine. Thus, in clonal propagation by nodal explants, there is no need for further re-inoculation. The bacterium stimulates plant growth and induces changes that lead to better water management. It also enhances resistance to low level pathogens of potato and tomato and to Botrytis cinerea infection of grapevine.

The inoculated plantlets had a massive and wellbranched root system after four weeks in culture, and were more advanced developmentally. The stems were sturdier with more lignin deposits around the vascular system, more root hairs and more and larger leaf hairs. Stomata function of the bacterized in vitro plantlets closely resembled those of the greenhouse hardened non-bacterized transplants. In wheat seedlings exposed to osmotic stress, improvement of water relations has been observed by their co-culture with Azospirillum brasiliense strain SP245. Enhanced transplant survival of black locust, inoculated with different Rhizobium isolates selected for growth vigour from Rabinia varieties, has also been reported.

The stimulation of adventitious rooting in cultures with Agrobacterium rhizogenes is well known. It was successfully used to enhance rooting of two Mexican species of Pinus. Rooting of somatic embryo-derived adventitious shoots (15-30 mm long) increased between 7-13 per cent, and in shoots co-cultured with A. rhizogenes to 67 per cent compared with 60 per cent in the auxin treated controls. Root development was also induced in slash pine explants on co-culture with the bacterium isolates. The bacterium-induced roots resembled seedling roots rather than hairy roots, which are induced with A. rhizogenes.

Induction of root elongation has been observed in potato plants co-cultured in vitro with some strains of Rhizobium leguminosarum biovar trifolii, isolated from the root nodules of red clover. The bacterization of shoot explants of oregano with a Pseudomonas sp. prevented vitrification. The oregano plantlets co-cultured with the bacterium had lower water content, but more phenolics and chlorophyll than the non-bacterized controls. Similar responses were recorded with potato and vegetable crops. Bacterized plantlets were greener, had elevated levels of cytokinins, PAL (phenylalanine ammonia-lyase), free phenolics (especially chlorogenic and caffeic acids), and contained more lignin. In potato, elevated PAL activity in bacterized plants was found during the first four weeks of growth, there being no difference in its level in six week-old plants.

However, when explants were taken from six-week-old bacterized and control stock plants, PAL activity in the newly derived shoots was higher in the pre-bacterized treatment. This suggests that pre-bacterized plantlets respond faster to cutting injuries, a critical factor in the induced host resistance to pathogen attack. In vitro bacterization of potato and vegetable plantlets significantly enhanced their post transplanting survival vigour. Enhanced vigour of potato plantlets co-cultured with isolates of Pseudomonas fluorescence isolated from tubers, during and after weaning has also been reported.

In greenhouse experiments, plants derived from dual cultures of potato and our pseudomonad bacterium had larger root systems and stolons, tuberized earlier, and gave higher tuber yields than the non-bacterized controls. The bacterized potato plantlets transplanted directly from culture vessels to the field had significantly better survival than the non-bacterized controls; however, the tuber yield varied from year to year; high precipitation or severe drought reduced the benefits of bacterization.

Tissue culture techniques provide an opportunity for the introduction of nitrogen fixing endophytes (other than Rhizobia) into clonally propagated plants for sustainable production systems. A protocol has been developed to inoculate micropropagated sugarcane plantlets with Acetobacter diazotrophicus at the end of the rooting period on medium with 10X diluted salts and sugar but without hormones and vitamins. Thirty days after transplanting, the bacteria could be re-isolated from the plant tissues. Stable artificial associations between plants and nitrogen fixing bacteria have been also reported. A symbiotic culture system of callus and bacteria between Daucus carota L. and Azotobacter zettuovii (CRS-H6) was developed.

The callus grew for four years on the nitrogen-free medium with lactose as the carbon source. The bacteria located in the intracellular spaces were transmitted to newly regenerated plantlets and fixed nitrogen. Similar stable associations were established with tomato, potato, wheat, sugarcane and poplar, using different strains of eight Azotobacter species (Varga, S.S. Personal communication), and in strawberries with Azomonas insignis. In tomato, the bacterium was transmitted through the seed. Approximately, 20 per cent of the seedlings from the seed of tomato cocultured with Azotobacter beijerinckii showed N2 fixing activity two months after germination.

Ex Vitro Bacterization

To comply with the certification requirements of plant tissue culture propagules, commercial laboratories are hesitant to introduce microorganisms into their in vitro cultures. One approach is to wean plants in rhizosphere with both beneficial bacteria and mycorrhizal fungi. In potato, the tissue-cultured plantlets are readily infected both in external and internal tissues with bacteria present in the transplanting medium. This approach, although attractive, will have to deal with the problem of maintaining stable populations of the introduced microorganisms.

The beneficial effects of bacterial inoculation on tree seedlings and greenhouse produced potato tubers show the potential of post-vitro bacterization of tissue-cultured propagules. The post-vitro mycorrhization and bacterization of micropropagated strawberry, potato and azalea with certain combinations of bacteria and mycorrhiza enhanced greenhouse production of minitubers, and a mixture of three strains of rhizobacteria improved the post-transplanting performance of strawberries.

Mycorrhization

The inoculation of tissue-cultured plantlets with endomycorrhizal fungi is beneficial to plant growth. It is well known that the root colonization with vesicular-arbuscular mycorrhizae (VAM) improves plant nutritional status, water management and disease resistance. Several secondary metabolites (cyclohexenone derivatives) with structural similarity to abscisic acid have been identified in tobacco roots in response to the Glomus intraradices infection. A variety of phenylpropanoid compounds induced by ectomycorrhizal inoculants, have also been identified in several conifers. Such compounds are related to the biotic and abiotic stress management in plants.

To improve the performance of tissue-cultured plantlets, rooted-shoots have been inoculated with mycorrhiza at transplanting. The benefits of mycorrhization depended on the growing medium, plant and mycorrhizal species, and the degree of root colonization. Because VAM stimulated growth of plants was frequently expressed only after acclimatization, it was suggested to develop a culture system where the mycorrhizal fungi are introduced in vitro during the rooting stage.

It is possible to grow fungi in vitro on Ri T-DNA transformed carrot roots under elevated CO_2. Capitalizing on this technique, a 'tripartite' culture system for in vitro inoculation of strawberries and vegetable crops was developed. The most effective treatment consisted of 30 day-old VAM transformed carrot-root culture and strawberry shoots in cellulose plugs. After root induction, the plugs were placed on the surface of the mycorrhized root culture and kept in a growth chamber under 5000 PPM CO_2 for 20 days. All plantlets were successfully colonized, and exhibited a larger root system, better shoot growth, and a higher (more negative) osmotic potential when compared to non-mycorrhized controls. It is suggested that the enhancement of osmotic potential is important in the pre-adaptation step prior to full acclimation of plantlets for transplanting. The high osmotic potential of wheat seedlings co-cultured with Azospirillum allowed them to withstand osmotic stress much better than non-inoculated controls with low osmotic potential.

In strawberry, mycorrhized plants had a better establishment rate and produced more runners than non-mycorrhized controls. The plants were grown in polyurethane foam substrate under photoautotrophic conditions with reduced sucrose concentration in the medium. The foam tear-away strips fit into any culture vessel, and the VAM spores can be placed directly in the planting holes. In garlic, improved growth was observed after post-vitro transplant inoculation. Increased rooting and reduction in weaning stress have been reported in the medicinal plant, Babtista tinctoria (L.) R. BR..

Several studies have demonstrated synergistic effects of VAM fungi and diazotrophic bacteria on nutrition and growth of various crops. Such effects of "mycorrhiza helper bacteria" need to be explored in tissue culture systems. Creation of defined tissue culture micro-ecosystems could allow the study of complexed plant–microbial-environment interactions, which could refine and improve our traditional in vitro propagation methods and prime the propagules for ex-vitro environments.

Priming with Microbial Derivatives

Cyanobacterial and fungal culture extracts have been shown to induce developmental changes and pathogen resistance in tissue-cultured material when added to the medium. Microbial culture filtrates have been reported for the control of plant tissue culture contaminants. An acetone precipitated fraction of Bacillus subtilis Ehrnberg and Trichoderma viride Pers exhibited high antifungal activity, and extracts from Pseudomonas fluorescence Migula (strain X) had high antibacterial activity.

The bacterial extracts did not reduce the performance of cultures of Nicotiana tabacum L. over four subcultures but reduced the growth of accidental contaminants. Heat-stable mycelial extracts of another Trichoderma species, a non-pathogenic T. longibrachiatum were triggered induced-resistance response to Phytophthora parasitica var. nicotianae (race 0) in tobacco seedlings. The extract induced expression of pathogenesis-related proteins, PR-16 and osmotin (PR-5) at a higher level than the extract prepared in the same manner from the pathogenic fungus.

MICROPROPAGATION PROTOCOLS

Micropropagation protocols have been developed for a large number of species; however, only a limited number of plants are being produced on a large scale through micropropagation. This is because conventional methods of propagation are cheaper than tissue culture technology or there is a selective market demand. Ornamentals account for approximately 80 per cent of the world trade in the tissue culture industry. Plants produced by commercial laboratories should yield steady revenue, e.g. Spathiphyllum, Syngonium, lilies. Some plants are unique to individual commercial companies. The life cycle of the species and the cropping system adopted by growers are important criteria to select plants for large-scale tissue culture.

Initially, micropropagation units focused on the production of ornamental plants that were least affected by the season, e.g. foliage and indoor plants. But to capture larger markets, commercial tissue culture units have ventured into fruits, vegetables, landscape and medicinal plants, which provide options for production around the year. This leads to improved cash flow and optimal use of equipment and facilities. When selecting specific varieties protected under Plant Breeders Rights, these should be multiplied only with the agreement of the breeders.

Target markets

Before going commercial, one must ensure that there is a large and expanding market for the product. Both the domestic and export markets should be explored extensively. It must be ensured that large contracts are available each year. Products that command a high price should be preferably chosen.

Regularity of Dispatches

It is essential to plan the production schedule according to the customer's demand and despatch the products at the stipulated time. It is not in the interest of the tissue culture facility to remain idle. Production of different plants should be distributed over the year to make optimum use of the laboratory facilities. Regularity of despatches adds to the viability of the business.

End Product

It is essential to decide on the type of products to be produced in a tissue culture facility. These could be plants that are rooted or non-rooted and could be sold in agar or marketed as hardened propagules in plugs or trays. Relatively small tissue culture units can produce nonrooted microcuttings in large numbers. Rooted microcuttings on the other hand require large growth room areas.

In both cases, dispatching them ex-agar may result in some mortality at the grower's end. To reduce mortality, it is important to ship them in-agar. However, in-agar shipments occupy a large volume, and hence require more cargo space and add to the cost. Such high transport and cargo charges can only be justified for high value products. The hardened micropropagated plants in plugs or trays are the easiest to handle by growers.

Prevention of Variation in Tissue Culture

The establishment of mother cultures from explants is a significant part of the production cost. Production projections can get out of gear due to the lack of sufficient mother cultures and maintenance of large numbers of such cultures adds to the cost of production. There is a general consensus that the number of subcultures for explants should be limited to avoid somaclonal variation. For example, in potato, the number of subcultures is recommended to be around twenty. Hence, repeated explant culture initiation becomes necessary to have sufficient number of mother cultures whenever largescale production is planned.

For many plants, it is not known whether limiting the number of subcultures reduces genetic variation. Many of the plants that are propagated from vegetative parts such as banana, potato, sweet potato, sugarcane, chrysanthemum and many fruits are polyploids and well buffered against the expression of mutations and somaclonal variation. If axillary bud proliferation is used as the strategy of multiplication, the frequency of variation is low. If variation does occur, the number of variants will rise with increasing number of subcultures unless there is selective elimination in tissue culture. The high frequency of somaclonal variation has prevented the use of tissue culture for mass scale propagation of oil palm. In general, the initiation of cultures from meristem-tip gives low variation in subcultures. Some of the novel somaclonal variants can be of value in expanding the market.

Multiplication and Hardening

Liquid media reduce the costs of multiplication. Increasing the multiplication rate through enhanced axillary bud proliferation helps in cutting cost of production. Mechanical cutting and bioreactors is a valuable option at Stage II. In some systems, shoot elongation and rooting are two separate phases. Combining them in one step by using appropriate media with auxins and cytokinins is a valid cost cutting strategy. Reducing the duration in the hardening facility but increasing the survival rate reduces production costs.

If natural light is used during multiplication and rooting, the plants need relatively shorter duration for hardening. Acclimatization in vitro helps in better survival when the plants are transferred to soil. Loss of plants at the hardening stage contributes to increased costs of production. A cost-effective system based on natural sunlight for in vitro acclimatization gives high survival. If grown under artificial lighting, tissue cultured plants should be kept under shade for some time and only then subjected to full sun hardening.

Greenhouse Facility

Sophisticated and costly greenhouses can be substituted by cost-effective techniques for hardening in vitro derived plantlets. For this, it is necessary to understand the changes that tissue-cultured plantlets go through during the transfer from in vitro to in vivo conditions.

Table: Changes in the Environment for Hardening in Vitro Plants to Ex-vitro Conditions

In vitro conditions	*Ex-vitro conditions*
Temperature 25 ± 2°C	Temperature 23-36°C
Light intensity 1200-2000 lux	Light intensity 4000-12000 lux
Narrow spectrum light	Broad spectrum light
Relative humidity 98-100%	Relative humidity 40-80%
Almost non-functional roots	Functional roots
Almost non-functional photo - synthetic system	Functional photosynthetic - system
Exogenous hormones	Endogenous hormones
Sterile conditions	Non-sterile conditions

Micropropagated plantlets should be acclimatized gradually by transferring them to a clean area under partial shade. In a low-cost hardening system, the culture containers are kept in the greenhouse with lids loosened. The plants are allowed to grow for 4-6 weeks under conventional Grow lights TM. This semi-hardens plants, and leads to shoot elongation, which is usually done under laboratory conditions. Contamination can be controlled with regular sprays of fungicides and bactericides.

The plants are periodically sprayed with water to maintain humidity. Bench lamps are used to provide light, if and when required. Semihardened elongated shoot clumps are then separated under non-sterile conditions, and rooted in peat or plugs under thin plastic sheets for 2-3 weeks. This low-cost alternative reduces the capital cost of hardening by saving laboratory space, labour, and time. This procedure has been used on a commercial scale for a number of plants such as Gerbera, Spathiphyllum, Syngonium, African violets, Chrysanthemum, Ficus, Philodendron and other ornamentals in the Netherlands and Israel. The growers in Florida harden tissue-cultured foliage plants in cavity trays covered with a transparent plastic tray, thus saving costs of a greenhouse. In some parts of India, hanging the culture plastic bags in shade hardens tissue-cultured sugarcane plants. Plants produced in this manner are semihardened, and can be delivered to the growers at a relatively cheaper price.

Low-cost plastic tunnels (polytunnel) with bio-fertilization can also be used for hardening, especially in cooler climates. A rectangular pit of desired and manageable size is dug on a site free of water logging. A frame of bamboo or any other available material is made above the pit with a gentle slope to one side. The frame is then covered with a clear polythene sheet, and tied using nylon ropes. The sheet is sealed with mud on three sides, leaving one side free.

The temperature and humidity inside these structures is generally higher than the outside, and protects the plants from frost. The polythene sheet can be partly opened to allow air circulation and sunlight as and when required. The added advantage of polypits is the carbon dioxide fertilization effect and reduced need for watering. The International Institute for Tropical Agriculture, Ibadan, Nigeria has developed another low cost hardening alternative. The institute provides the farmers with plants in plastic bags and handling instructions. For hardening the plants, the farmers hang the plastic bags under tree shade.

CLONAL PROPAGATION OR MICROPROPAGATION

Genetically identical plants derived from an individual are called clones. Processes that produce clones can be put under the term 'cloning'. This includes all the methods of vegetative propagation such as cutting, layering, and grafting. Propagation by tissue culture also helps in producing clones. Using the shoot tip, it is possible to obtain a large number of plantlets. This technique is used extensively in the commercial field for micropropagation of ornamental plants like chrysanthemum, gladiolus, etc. and also crops such as sugar cane, tapioca, and potato. Thus an unlimited number of plants that are genetically similar or are clones can be produced in a short span of time by tissue culture. It has been demonstrated that a variety of plant species can be conveniently propagated through the techniques of cell, tissue or organ culture. This is popularly described as clonal propagation ormicropropagation.

The major benefits of this method include the following:

- Rapid multiplication of superior clones and maintenance of uniformity;
- Multiplication of disease free plants
- Multiplication of sexually derived sterile hybrids.

In most cases, clonal propagation is achieved by placing sterilized shoot tips or auxillary buds onto a culture medium that is sufficient to induce formation of multiple buds.

This method has already been used to propagate a large number of marketable ornamentals.

Following stages are involved in the method of clonal propagation:

- Stage I involves establishment of tissue in vitro.
- Stage II involves multiplication of shoots (often media are not changed between stage I and stage II).
- Stage III concerns root formation and conditioning of propagules prior to transfer to the green house; this stage requires alteration of media for promotion of root formation.
- Stage IV involves growth in pots followed by field trials. The number of steps in micropropagation may sometimes be reduced to three or two as shown in diagram.

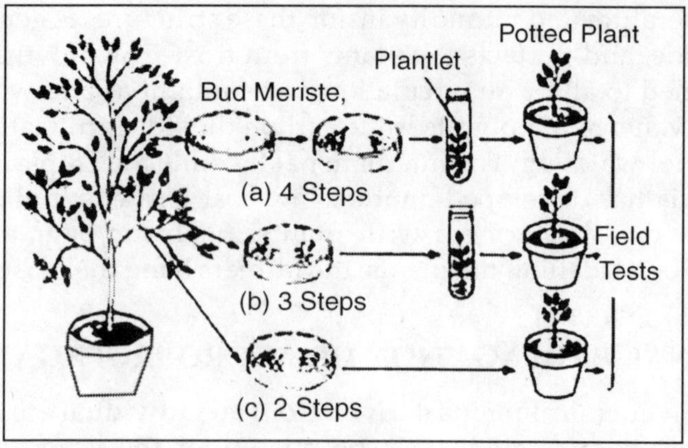

Fig. Three Different Methods of Micropropagation of Plant Material

A wide range of plants have now been regenerated through technique of tissue culture. This has been found particularly useful for propagation of tree species, so that a large number of plant species have been successfully grown by tissue culture.

LARGE-SCALE PROPAGATION

To bridge the gap between research and application, the Department of Biotechnology, Government of India sponsored the setting-up of two pilot-scale facilities for large-scale propagation of elite planting material of forest trees through tissue culture. One of these facilities has been established at TERI's 36-hectare-campus in Gual Pahari, Haryana with an annual capacity of a million plantlets. Research at these facilities focuses exclusively on developing new protocols for mass cloning of elite planting material, mainly of trees.

Till date, over 4 million plants have been dispatched for field plantation from these facilities. The tissue culture raised plants are presently being evaluated under field conditions. This is being done in tandem with the forest departments of Haryana, Uttar Pradesh, Madhya Pradesh, Bihar, Jammu and Kashmir, and Orissa. For initial screening for phenotypically superior trees only a few hundred plantlets of the same are raised and tested under various agroclimatic zones. The best clones are then mass multiplied and monitored regularly for their performance. Field data suggest a survival percentage of more than 90% even in the harsh conditions of Aravalis without the life-saving irrigation. At half the rotation age some of the selected clones of Eucalyptus are showing a significant increase in productivity as compared to the conventional seed raised progenies.

ADVANTAGES OF CLONAL PROPAGATION

Clonal forestry is often seen as modern biotechnology. It is true that some techniques used for creating clones are new. The technical capacity to produce clones is improving. The current paper aims to review the possible advantages of clones in forestry and forest tree breeding and to give information about the occurrence of clones in nature and forestry. Some specific attention is given to focus on matters, which are relevant for cloning as a tool to get quality birch. No effort is done to discuss possible disadvantages and problems in full, but this has been done many times before. Neither are the possibilities of other "competing" alternatives to get some of the advantages (like using controlled crosses) fully explored.

Clones in Nature and Forestry

Clones are common in Nature, and they have been used as a tool of domestication for very long time and are used in large-scale forest operations. It seems to be important to remember that to put clonal forestry into perspective. The earliest land plants lacked vascular tissue and had inefficient systems of sexual reproduction and dispersal and clonal reproduction was the typical way of propagation in nature. Clonal growth was advantageous enabling physical dominance of large areas through horizontal growth.

Clonal propagation is still important in nature for many species. Many of the commonest clones are herbaceous but the habit is also found in woody plants throughout the world. In temperate broad-leaved woodlands, root suckering is a common form of clonal growth and there are good examples in the genera Populus and Prunus. Coppice is a natural growth form that has been exploited for centuries in European forests. Betula, Carpinus, Corylus, Quercus, Salix and Tilia are all genera with the ability to self-coppice. The use of vegetative propagation of trees as a tool in their domestication has a long history. An early use was to propagate good clones of fruit trees and this has been done for thousands of years, thus humanity has long experiences of cloning trees.

For specific cases clonal forestry is often regarded as a "new technique", and "biotech" leads associations to the public to something more radical than most cases of clonal forestry. To put things into perspective it has to be remembered that many of the techniques used in modern "seedling forestry" are also "new", and the problems and uncertainties with that is often larger than the uncertainties connected to clones.

Sugi

The first known use of vegetative propagation for forestry purposes was in Japan with Sugi in the 15th century. Since then reforestation with monoclones or clonal mixtures has been widely used in Japan. Cryptomeria japonica is the most widely cultivated tree species in Japan comprising nearly half of the ten million ha of Japanese plantations. Almost half of the plants used are clones. The extreme uniformity of the plantations has probably contributed to greater than average damage from typhoons and heavy snow.

Attacks from Sugi bark borer and bark midge have increased in severity as the area of Sugi plantation has grown, but there is no evidence of altered disease or insect virulence related to clonal forestry, despite careful monitoring.

Clonal forestry is most common on the south-western island of Kyushu where about 100 cultivars are in use. Many foresters use only 1-3 cultivars of local origin in their plantations so that a single clone may cover as much as 2 hectares. Allergy to Sugi pollen is a health problem in Japan as a whole, but is only a minor problem on Kyushu with its large use of clones. Cloning by cuttings for several generations has reduced the incidence of male flowers, so these plantations produce little pollen. Analogously one may speculate in that a reversion of North European birch forestry to clonal forestry in a very remote future may reduce allergic reactions with birch pollen.

Eucalypts

Clonal forestry is more common with broadleaf species than with conifers. Probably the most successful example is the eucalypt plantations in countries in tropical and subtropical climates like South America, South Africa and Portugal. The total area of eucalypt plantations today is about 30 million hectares and about half of this area is planted with cloned material. The eucalypt plantations in Brazil cover an area of approximately 3 million hectares.

A succession of eucalypt species has been grown in Brazil due primarily to disease problems. Clonal forestry started in the late 1960:s and is today the dominant form of eucalypt forestry in the country. Progeny from an Eucalyptus grandis female (male sterile) growing in an arboretum were found to be natural hybrids which were resistant to pests and diseases. The offspring of this tree are believed to be Eucalyptus grandis ′ Europhylla hybrids. This knowledge was rapidly adopted by the forest industry companies and they selected clones within the open-pollinated offspring from this single tree, which means that a small number of clones originating from the same family are widely grown over large areas in Brazil.

This has and is still changing with most companies having good breeding programmes backing their clonal programmes. The clones are deployed in huge monoclonal blocks. This situation seems favourable for a pathogen, but nothing has happened yet - despite the possible risks of a disease in a monoculture.

Today the companies have become more risk-conscious and are developing breeding and clonal testing programmes that include strategies for clonal diversity in space and time. A typical programme is now deploying about 20 clones each year and these clones will be replaced with time as new clones are selected. The production landscape forms a genetic mosaic. New genetic material has been imported to widen the genetic base for breeding.

Poplars and Willows in Europe

Vegetative propagation of poplars (Populus sp.) has a long history in Europe. Organised clonal forestry started in the beginning of the 20th century. Monoclonal plantations of poplars have become a common land use especially on river plains in southern European countries like Italy, Spain and France. Some of the best poplar clones in use were selected as early as during the first half of the 20th century. Clonal forestry with poplars is common in countries with subtropical and temperate climates like Italy, Spain, France, Belgium, USA and Canada. Single clones of poplars have been propagated extensively. There have been disease problems and the overuse of single clones and use of large monoclonal plantations has been questioned.

Poplar has covered over 100000 ha in one country and a single clone may comprise one third of this area. The focus has been to develop disease resistance within individual clones, but these clones have proved to be highly susceptible to new varieties of disease. The disease problem seems to be growing and has caused Germany to more or less abandon poplars and other countries may follow.

Nevertheless these incidents have not provided poplar growers with sufficient incitement to focus on more diverse alternatives. It is unlikely that monoclonal cultures and the wide use of some clones are the only explanation to increased disease, but they act as contributing factors. Legal and commercial reasons often favour the use of single clones.

Even willows and short rotation forestry in Sweden are threatened by disease. Resistance is broken down within a single rotation (it is decades in between replanting even when harvesting is done at intervals of a few years). Two evaluations of the Swedish willow breeding programme have rather strongly recommended an increased number of selections to be made (more than a single selection released per year). Nevertheless, the growers' organisations still do not encourage clone mixtures.

Hybrid Aspen in Finland

A Finnish forest industry company has recently started a clonal forestry programme with hybrid aspen. This raises the question if something similar could be done with birch. It may be noted, however, that suggested that micro propagation costs are high and that the technique needs development.

Overoptimistic Statistics

It is a clear tendency that statistics reported about the use of vegetative propagation, at least there it is regarded as modern hitech biotech, gives the impression of more rapid progress than actually occurs. It is reported, based on a survey, that four Swedish operators produces 7.3 millions rooted Norway spruce cuttings annually. It is estimated the annual production to 5 millions and believed in a raise to 10 millions. However, looking backwards, less than

20 million were planted in total till 2000 and now only some hundred thousands are planted annually.

Crop Uniformity

As genetic diversity and possible uniformity is an evident and possible important aspect of clonal forestry. For evaluation of possible ecological impact diversity seems central. I will thus start to discuss disadvantages with a low genetic diversity in a stand. In other respects this paper focuses on the advantages of clonal forestry.

Disadvantages of a Uniform Crop

There are reasons to believe that the biological production can be higher in a genetically diverse crop than in a uniform:

- A single genotype demands the same things at the same time, and thus utilises the site worse than a mixture of genotypes.
- In a mix, another genotype may take over the ecological space left by a failed genotype.
- A disease spreads faster in a uniform crop.

These statements are supported by the average of a very large number of agricultural mixing experiments and a few experiments and experiences with forestry. On average, mixes performed some per cent better and were less susceptible and more stable than the average of its components, but usually the production of the mix was not better than its best component grown pure.

Even if mixes seem more productive on an average, there are many examples and experiments, when no advantage was found and some experiments where diversity actually seems negative for production. I conclude that from a production point of view an advantage with some degree of diversity is expected, but that does not imply that production is higher the higher the diversity is. But the superiority in biological production is not expected to be large and it is not certain for individual cases, but just a rough generalisation. This general statement ought to apply to birch also, thus biological production is assumed to be lower and more uncertain in a uniform crop, but the effect is probably small and practically unimportant (a few per cent of biological production).

Advantages of a Uniform Crop

There must be considerable advantages in uniform monoclonal blocks as they are used frequently. Although the advantages with diversity are repeated and reiterated and very politically correct, it is seldom applied in intensive agriculture or husbandry. The main reason is that maximizing biological production and maximizing economical production is not the same thing. It makes it easier to manage the crop if it behaves uniformly. For a forester it may not matter very much if individual birches drop their leaves on different days, but a farmer wants the crop to become ripe the same day, as the whole crop is harvested the same day. Diversity is a usually a problem for the customer, who wants a uniform raw material for further processing. For administrative processes, like describing the crop, it is an advantage if it is uniform and does not change from specification over time and space. These problems are evident for many types of agricultural crops, and therefore these crops are usually uniform even when biological production would be higher with a diverse crop.

The dream of many foresters is to find the best genotype and use only that. This has been successfully done for other domesticated species. The uniformity itself in a clonal plantation has a considerable advantage from a management point of view. It is easier in the nursery to raise plants of a single genotype. It is easier to manage and harvest a uniform forest. It has advantages to handle a single type of material. A customer or processor of a product knows more about a lot if it is a specified clone than if just the species is known. A batch of uniform logs is worth more and easier to handle than the same amount of very variable logs. Even if interest is only in quantitative production and that is lower in a uniform crop, still the higher genetic performance of the best clones compared to the best seedlings available usually much more than compensates for the production loss by a uniform crop. Thus uniform clonal crops produce more than diverse seedling crops.

My opinion is that genetic diversity often is overemphasised in comparison with the higher genetic gain achievable at the cost of a reduction in genetic diversity for future forest production plantations in many countries (like Sweden) both for propagation with seedlings and cuttings. To get genetic diversity many clones are used in seed orchards and clonal mixtures and there constraints of relatedness of clones may be seen as desirable. This was not a disadvantage at an early stage of forest tree improvement, as it does not influence gain much if many phenotypically selected plus trees are placed in seed orchards, and in the early stages of cutting propagation there were physical constraints on the number of ramets which can be produced by each clone, so it is necessary to have many clones. But now this constraint often means selection is not intensive and the best genotypes are not exploited to the extent, which would be possible.

However, clonal forestry is unlikely to deal with a single clone only. That does not seem a likely scenario for birch. Many clones will be tested and test results as well as experiences will accumulate of the best clones, so the opinion of what is the best clone will change over time. It is also dependent on the locality, the desires of the user and the availability. Even if uniform plantations are planted, there will usually be a mosaic of clones on the landscape level. From the risk management point of view a clonal deployment philosophy offers advantages.

If a clone is sensitive to a pest or pathogen the problem can be identified and managed. If silvicultural means to deal with a problem are insufficient, the stand can be salvage harvested and replanted. The clone which turns out inferior can quickly be taken out of production. It is an advantage for an operator which uses monoclone blocks to use several clones in the same time that it will become more evident what the strong and weak point of the clones are. A disadvantage with mosaics is that the uniform lots of wood from a stand are often too small to be practical as a specific variety at the end user, and actually the end user may get trouble with larger between lot differences if clonal forestry is used. Alternatively the clones can be grown in intimate mixtures, in that they most of the possible advantages of the diversity of seedlings may be gained by a mix of rather few clones.

When we are talking quality birch - and not quantity birch - the reasons for monoclonal plantings become stronger. When high quality of the product is concerned, it often becomes more important to have uniform quality also, while in a low quality product - as firewood - the quantity is usually relatively more important and a uniform and predictable product is less important. Attention on quality instead of quantity also generally coincides with a higher level of intensity, and at a higher level of intensity, clonal forestry often becomes more interesting. Thus considering the aim the project arranging this symposium has in mind; uniform clonal plantings becomes relatively more important than for forestry in general.

Improved Clones can be Preferable to Seedlings

There are a number of arguments for forestry with clones. It takes less time to multiply a genotype as a clone than to build a seed production unit with that clone as one of the parents and wait till that works. The time lag is actually much longer than till the first seed production, because a seed production unit is usually designed to last for many years. The risk for getting the seed storage empty means that forestry often want to have five to ten years foreseen seed need in storage. That is expensive, many seeds may never be used and the seeds actually used are less improved than possible.

A seed production unit is designed to meet a certain need of plants with certain characteristics many years ahead. If too many seeds are produced it means that the investment was unnecessary large. If too few seeds are produced it may mean that foresters must use genetically inferior unimproved seeds. If the propagation is done fast by vegetative propagation by e.g. micro propagated clones or somatic embryogenesis, it takes only a year to multiply a desirable genotype in the desired number of copies. Thus it is fairly easy to adapt to the latest news from breeders and customers. There are large annual variations in seed production both in the forest and in seed orchards. In a seed orchard the reproductive output is variable among genotypes and the reproductive phenology may make clones incompatible, so the actual genetic set up of the progeny is a bit uncertain. These uncertainties can be avoided by clonal forestry. Many of our forest trees - like birch - can both spread pollen and produce seeds. That means that selfing can occur and the occurrence of selfing in a plant material will decrease the forest production. With the use of clonal forestry selfing can be completely eliminated.

In a wind pollinated seed orchard wild uncontrolled and unimproved pollen may contribute to pollination, clonal forestry is a way to eliminate this problem. However, for birch, seed orchards are small enough to be managed in plastic green houses, where the inflow of unimproved pollen is low, so for birch it is not fair to see this pollen contamination problem as an argument against open pollinated seed orchards.

In a seed orchard it is important to avoid related clones, as relatives get inbred progeny when they mate, which means that forest production drops because of inbreeding depression. Mating between relatives may be a worse problem than selfing in future seed orchards, as self-zygotes usually die; self-seedlings are often culled before planting; or early out-competed in forest, while milder inbreeding is more directly transferred to production loss. If clones are related in a clonal forest, that does not cause inbreeding depression, and thus there is little harm if related clones occur in a stand (actually this is very natural, trees in a natural stands are often related). The clones in the breeding population will be increasingly more related as breeding goes on, it will be easier to handle the consequences of this for the production forest with clonal forestry.

A seed orchard is a big investment intended to meet the seed need for a long period for a specific species used in a specified area, thus a considerable local market is needed to justify the investment. Conifer seed orchards are usually designed for a need of many millions plants per year over some decades. Multiplication of a clone can be seen as a smaller investment during a shorter period, thus the local market is allowed to be much smaller with clonal forestry, e.g. micro propagation for curly birch in Finland serves a market of 200 000 seedlings a year. A seed orchard has a stiff structure, its composition can be only slightly modified once it is established and it is not easy to justify several orchards heading for different characteristics for the same area. Clonal propagation is more flexible; an individual customer can choose the clones, which fits the particular needs for the customer and occasion. Sometimes a combination of rather rare characters is desired, a clone can be found which combines several desirable characters in a much more efficient way than dealing with seed orchards.

It is often stated that clonal forestry reduces genetic diversity. This is a generalisation, which need not always be true. Actually, clonal forestry provides a tool to choose genetic diversity at will, while when a seedling material is used, the genetic diversity is essentially left to chance. Probably those using intensively managed birch plantation for producing high quality timber with special characteristics would choose the option to reduce or eliminate genetic diversity on the stand level, because they found it economically sound to do that. It seems likely this is how the tool will be used. But clones in a clonal mix can be chosen so that the genetic diversity is larger than in a seedling material.

Clones can be mixed with seedlings, that may be a good strategy not only for diversity, but economically in cases where cloned plants are much better, but also much more expensive. If the clones succeed well, their share at the final harvest will be larger than at establishment. Over time and space different clones will be used even in a system with monoclone cultures (like Eucalypts in Brazil). In such a system the genetic diversity among stands is likely to be much than in nature. Sometimes clones have been used for natural conservation in situation when good seeds could not be obtained. Clonal forestry may strengthen the trend to produce the needed wood at a limited acreage of intensively managed production forests, allowing larger areas to be set aside for other purposes. Thus the question how clones may affect diversity has many aspects and it is an oversimplification to state that it must be severely reduced. A common situation is that some good seeds exists or can be obtained, but quantitatively insufficient for the plant need. If such a situation occurs unexpected and unforeseen, vegetative propagation may be the solution. A more common situation is that a few superior seeds can be produced by controlled crosses, but too few to meet the plant demand. Clonal propagation is often a way to make forestry with controlled crosses practical feasible. Clonal propagation may be seen as a way to amplify the impact of a few good parents or a tested cross by mating the parents and when multiply the seeds.

The genetic value of a genotype depends on the breeding value of its parents, but also how well the genes of the parents fit together. This may be called dominance. The later reason for genetic differences cannot be exploited in an ordinary seed orchard, but it can be exploited with clonal forestry. Hybrids sometimes result in good seeds. It may be difficult to make many of those seeds by controlled crosses, but hybrid seeds (or trees) can be multiplied by vegetative propagation. This is used in the Eucalypts and Aspen programmes mentioned above.

Personally I am not that convinced that hybrid superiority is so common as it is often stated. When hybrids are better than the parents it may be because the parents were not well adapted for the conditions where they were tested or that the hybrid combines characters of the parents for the environment and purpose of the forester in a better way than tested alternatives. Hybrid maize is a best seller, but the reason is more that farmers are unable to produce hybrid seeds them self, but has to buy them, and when it is good economy of a seed company to invest in hybrids, so they can market the seeds. To get a genetically superior forest, it is desirable to use the genes of the very best clones. In a seed orchard it has to be rather many parents to avoid selfing and problems with lack of overlap in flowering time. This means a drop in breeding value. For clonal forestry the very best clone can be used. This is true even if in the complete absence of dominance.

As dominance is seldom the major cause of genetic variation for the most important characters in forest tree breeding, this more effective exploitation of breeding value is usually a more important reason for clonal forestry than the exploitation of dominance variation, at least as long as we do not talk about hybrids of different species or races. Quantitatively the low importance for the superiority of cloning of a limited amount of dominance variation is demonstrated.

We have also special features for clonal selection. We can intentionally select clones, which do not waist resources on sexual activities but spend resources on stem wood production. Such clones are desirable for forestry. In a seed orchard it is not desirable with clones, which do not produce seed and pollen, so such clones would not be chosen for a seed orchard, and if they were, they would be unable to transmit their genes to the seed crop. Cloning methods have different characteristics. They are often very depending on the state of the material which is cloned, cuttings usually do not work well on physiologically old material, for somatic embryogenesis other tissue than immature seeds are often a difficult starting material. With micro propagation it is usually possible to multiply old trees or clones, and this is often an advantage.

Cloning methods have different characteristics. They are often very depending on the state of the material which is cloned, cuttings usually do not work well on physiologically old material, for somatic embryogenesis other tissue than immature seeds are often a difficult starting material. With micro propagation it is usually possible to multiply old trees or clones, and this is often an advantage. Cloning and the cloning method can have effects on the physiological status of the tree. E.g. conifer grafts have not thick bark at the bottom of the trunk. The protocol for making clones can have effects on the plants. These effects sometimes look desirable and can be exploited. For example Norway spruce cuttings seems to suffer a lower mortality from pine weevil damage than seedlings.

Clones usually origin from older and more mature tissues than seedlings and this affects the characters of the plants and may have effects also on the mature trees. I guess these differences to seedlings are not dangerous and nursery raised seedlings can also be claimed to be unnatural, and they my be different because of different growing regimes. I suggest that we can intentionally interfere with the characters of the plants by the production system and that clonal forestry gives additional options to use this in a positive way. Many of the factors I have mentioned above will contribute to that clonal forestry gives a genetically more improved forest than comparable seedlings.

An example of the development of gain over time for the clonal option versus the seed orchard option is shown in Figure. This is based on a simulation of the Swedish Norway spruce (Picea abies K.) long term breeding programme with inputs chosen to correspond to the real programme as close as possible. Clonal forestry and seed orchards are different ways of harvesting gain from the breeding population and clonal forestry means that the genetic improvement reaches a certain level three decades before the same level can be achieved by seed orchards.

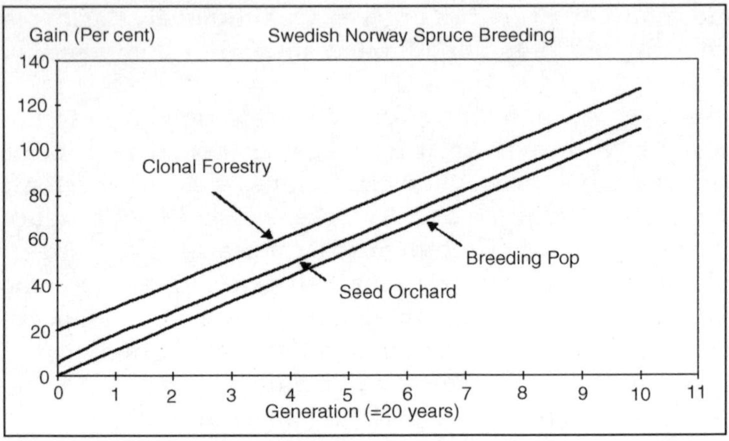

Fig. A comparison of the genetic value of three populations as a function of time based on a simulation of the Swedish Norway spruce breeding Programme.

Genetic progress is primarily created in the breeding population. The production population is created by creaming the best available material from the breeding population either by clonal forestry or seed orchards. The gain unit can roughly be interpreted as per cent gain.

The details of the figure depend on exact specifications, but it gives an idea of magnitudes based on best possible estimates. Clonal forestry gives a higher gain than seed orchards; the difference at a given time is in the magnitude of 15%. The same gain is achievable from seed orchards; it just takes a few decades longer to get it.

More Effective Breeding with Clones

It was pointed out that clonal forestry might be a faster way to better forests than seed orchards. Good genotypes are multiplied as grafts in most current forest seed orchards, thus clones are useful also for producing genetically improved seedlings. But there is another - still more powerful - way of using the clonal tool to improve the genetic quality of forests. Clone testing has the potential to make the long-term breeding considerable more efficient.

Even if clones are too expensive for direct use in forestry, their use in testing of genetic materials may help forestry to get better seeds in the future.

A combination of testing clones for forestry and for long-term breeding has synergistic effects, thus such a programme is more efficient than a programme focusing only on one of these two aspects. Higher organisms are diploid, thus have two set of homologous chromosomes and carry two packages of genes. The parents only transfer one package (a haploid gamete) to the progeny, and the composition of that gamete is different for each progeny. Thus, if offspring is used to test parents, only half of the genes in the offspring are relevant for the parent, and these genes will be different for each offspring individual. In contrast, in clonal testing all genes in the genotype are tested and no genes generate random noise in the statistical sense.

The performance of a clone depends on its father, its mother and how the genes of the mother and father match together, thus dominance effects. However, dominance variation does not seem large enough for most characters of practical interest to be important. Thus the performance of a clone in a clone test says much about what will be inherited to its progeny. The accuracy by which the breeding value of a clone is determined increases with the number of ramets.

The uncertainty caused by the environment decreases then environment is repeated. Thus clonal testing is a way of increasing the accuracy. One can see it as a way of increasing the heritability to decrease the environmental variance by letting the environment be the average of a number of individual environments, one for each ramet. The gain obtained by a genetic selection depends on the selection intensity and the accuracy of the test. Under the constraint of a fixed plant number, a lower number of genotypes can be tested if genotypes are replicated by cloning, and thus the selection intensity will decrease It thus becomes an optimization problem.

However, if the individual heritability is not very high and the resource budget is not extremely tight, the optimum is to clone in at least some copies. A seedling material is genetically diverse; the best genotypes grown pure are much better than the average of a seedling material. The performance of a seedling is mainly depending on its environment and not so much of its genes. If it is cloned, the average over the clone is more depending on the genes. Testing by clones can be seen as a way to eliminate the statistical noise. The heritability of clonal values will be much higher than the heritability of seedling values.

Each seedling is genetically unique; a scientific principle is to use replications when testing. Thus a seedling is a rather unsatisfactory material for genetic testing and conclusions based on seedlings will be dependent on model assumptions, which are uncertain and affected by statistical noise. Clone testing is less depending on model assumptions and a more direct measure and thus offers theoretical advantages! In a series of investigations the candidates for the breeding population by clonal testing instead of phenotypic testing or progeny testing, if that is at all possible.

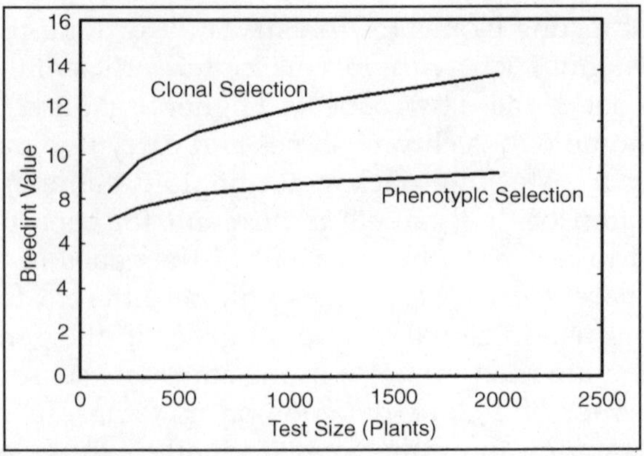

Fig. Comparison of Clonal or Seedling based Testing for the Swedish Norway Spruce long term Breeding programme (the X-axes gives the Resource level, the test Size for the real Programme is Around 560 plants/parent).

Clonal testing adds around 30% to the gain of the breeding programme, the advantage is larger the more resources that are available for each parent.

The superiority of clonal testing over phenotypic testing is demonstrated in Figure. It is based on the same type of simulations of the Swedish Norway spruce long term breeding programme as the previous figure, although it varies the resources available. The resources available today correspond to 560 test plants per parent. The more resources available, the more reason to use clone testing to get a more accurate selection (compared to increasing the number of test genotypes for a more intense selection).

Lstiburek and Danusevicius and Lindgren studied the conditions under which clonal testing is a favourable alternative. The results show that clone test is superior to phenotype test and also progeny test under a wide range of conditions. Thus even if the genotypes or ramets are expensive; and over a wide range of available resources; and in the presence of typical dominance variation; clonal testing can be recommended. In cases where the individual heritability is close to 1, clonal testing is less favourable. When the phenotype itself gives sufficient information about the genotype, the accuracy need not be improved by cloning. Such cases are rare for growth characters, but may be more common for some quality characters.

Numerous studies have shown that environment can cause lasting changes in phenotype that can be passed from one generation to the next, much as genes are transmitted. Such affects may limit the usefulness of clonal testing for breeding value, but similar objections could be made against progeny testing, and the effects must anyway be larger than usually observed (e.g. Lindgren and Wei 1994) to really matter.

Natural genotypes are not designed for perfect cooperation; the Darwinian selection in the struggle for life unavoidable gives some preference to strong competitors. A certain degree of selfishness is an unavoidable consequence of natural selection. By clonal selection we ought to be able to identify and utilise the best collaborators, who act in the interest of the stand and not only the individual tree..

A forest material is used over varying conditions in time and space and the forester wants a reasonable stable material over these variable conditions. Clones can be tested over time and space. That make breeders able to select clones, which perform well everywhere and anytime. Natural selection works just here and now. Again a case where Man may find more effective ways than Nature has used.

Tree Breeders first select based on field tests and when mate the selections. Artificial crosses are difficult to organise in the remote places where field tests are localised. But if clones are tested, some copies could be placed where it is convenient to make artificial mating, like in archives outside the institute. Foresters do not want trees to waist resources on sex, but to do breeding it is necessary to get reproduction to work. But clones, which have little reproductive output in the field, can have copies at the research station, which are manipulated to make mating possible. Thus with clones different individuals can be used for test and reproduction. Genetically modified trees will play a role in practical forestry and tree breeding at some time in the future. Such implementation is much easier, if clonal forestry is an established praxis.

INTEGRATION OF TECHNOLOGY FROM LAB TO LAND AMD PRICE OF THE TISSUE CULTURE

The price of the tissue culture derived plants, tubers, bulbs, and cuttings must be competitive with those obtained from conventional propagation. Many tissue-cultured plants are too expensive for direct field planting. In such cases, the cost can be reduced by one or more times conventional propagation of the tissue cultured plants. The uniformity and consistency in field performance of tissue-cultured plants is important to build confidence of the farmers to integrate such plants in the production systems

Tissue culture technology contributes significantly to the improvement of agricultural productivity and food security. However, it must be integrated into production systems in a cost effective manner. Wherever practical, direct use of the tissue-cultured plants is the best strategy to obtain the full benefits of this technology. The price of the products, i.e. plants, tubers, bulbs and cuttings, must be competitive with those obtained from conventional propagation. While many plants that are propagated from vegetative parts can be micropropagated, the cost of tissue-cultured plants is still too high for direct growing by the farmers.

For example, micro- and mini-potato tubers'and micropropagated plants of chrysanthemum, strawberry, sugarcane and cassava are too expensive for direct field planting. In such cases, tissue-cultured plants are grown as super-elite or elite material and the costs are brought down by conventional propagation one or more times. Thus, the costly tissue-cultured plants are used as nuclear material, and their further multiplication is by conventional vegetative propagation in the field to produce large quantities of pathogen-free planting material.

However, the subsequent vegetative multiplication of tissue culture-derived material in the open environment also exposes them to infection with diseases and viruses that may increase in successive propagation. In addition, the handling of bulky conventional planting material, such as potato tubers, banana corms and sugarcane sets adds to the transportation costs. Hence, the cost of tissue-cultured plants has to be brought down substantially, so that farmers benefit through increased productivity.

ON-FARM TECHNOLOGY TRANSFER

The most outstanding application of tissue culture in crops that are propagated from vegetative parts has been the production of disease-free planting material that gives improved yield and quality. This has been particularly the case in potato, banana, citrus, date palm, pineapple, papaya and many ornamentals. Seed potato production is a highly specialized business, and requires highly skilled growers.

Currently, all the certified seed potato in Ireland and Scotland is derived from tissue-cultured plants, which produce micro- or mini-tubers that form the super-elite seed. The mini-seed tubers are not used for growing the potato crop. Instead, these are further multiplied by conventional propagation in the field, at least twice, and sometimes even 3 to 4 times for the production of certified seed tubers.

Apart from cost reduction and high seed quality, the farmers get the same types of seed tubers with which they are used to. In Canada, many growers have integrated small tissue-culture laboratories for seed potato production on their farms. Similarly, the sugarcane growers in Punjab, India have switched to high quality seed cane that is derived from tissue cultured plants, which are propagated 3 to 4 times from conventional cuttings. The lead in commercial sugarcane tissue culture technology has been taken by the sugar cane mills, which have established their own laboratories for large-scale seed cane production.

The state agricultural university provides the disease-free indexed cultures to the sugar mills. Such a model that includes the university for research and trouble-shooting problems, private industry (sugar mills) that multiplies the sugarcane seed material through tissue culture, and cane growers who produce the crop, has been very successful in integrating tissue culture technology for the benefit of the farmers and processors. This tripartite cooperation has rapidly replaced conventional sugarcane seed production in the state. Recently, direct planting of micropropagated sugarcane plants has been profitably undertaken in Maharashtra, India. In Morocco, citrus indexed scions are multiplied in tissue culture and grown under a plastic house. Scion cuttings with two-three buds each are then sold to the farmers, who graft them on either new or old rootstock. This has helped to eradicate many of the citrus viruses and yet retain the old rootstock plants that are resistant to diseases. In another approach, high quality disease-free tissue-cultured scions are grafted in vitro onto seedling rootstock. Millions of such in vitro grafted plants have been distributed to the citrus growers in Spain, bringing direct benefits to farmers.

In Thailand, tissue-cultured chrysanthemum plants, produced by the Department of Agriculture, are sold to farmers who multiply them further from conventional cuttings for the production of cut flowers. This procedure has allowed rapid introduction of new varieties to respond to the market demand. In Vietnam and China, in vitro cultured plants of sweet potato and cassava are distributed directly to the farmers. The farmers multiply in vitro plants further from conventional cuttings, and produce sufficient amounts of healthy planting material for their on- farm use. A similar procedure could be used for the distribution of disease-free material of cassava in Africa where cassava mosaic virus disease has become a major threat to cassava production.

Perhaps, the most successful on-farm use of tissue culture technology has been in Cuba where an ordinary village-house is converted into a plant tissue culture facility – a 'biofactory'. Such a facility employs local labour, particularly housewives, for various operations, uses low cost media and containers and sunlight, and has the capacity to produce one million banana or sugarcane plants per year. In banana, the hardened plants are sold to the local banana growers. A batch of one thousand plants is kept in the nursery to monitor the quality. The farmers are asked to notify any off-type, which they may observe. This has not only given the farmers a confidence in the technology but has also resulted in the discovery of several improved variants among the tissue culture derived populations.

In developed countries, many commercial companies restrict tissue-culture propagation to the production of in vitro plants, which are used as nucleus or 'super elite' material, and are further multiplied two to three times before being handed over to the farmers as certified material. Other companies subcontract the production of fully-grown plants on a buy-back basis for further sale. This approach has been used in the propagation of many ornamental plants and shrubs in Belgium.

In the developing countries, tissue culture must be seen as a complementary approach to conventional plant propagation, so that the rural job opportunities are not lost, but strengthened and further expanded. Hence, on-farm conventional multiplication of the tissue culture propagated material is the cheapest and the most sustainable method for cost reduction, and for the creation of farmer confidence in the technology. Of course, the agricultural extension workers and propagation companies must provide the necessary know-how to the farmers for handling and growing tissue cultured plants for further multiplication under high health status. Universities and experts should provide training to farmers and extension workers as being done in several countries.

On-Farm Performance

Even after a decade of introduction of millions of tissue culture plants, the market covered is less than 1 per cent of the total cultivation under vegetatively propagated crops. Thus, there is a tremendous scope to use tissue-culture technology to provide growers with nuclear stocks of shrubs, fruit and forest trees. There is a need to micropropagate hundreds of native varieties and local cultivars of banana and plantain, cassava, citrus, date palm, ginger, potato, sweet potato, turmeric, yams, and local tropical fruits, flowers, herbs, medicinal plants and trees that are propagated from vegetative parts.

The consistency in field performance of tissue-cultured plants is extremely important. Farmer's participation is essential in achieving this goal. Precision agricultural management, based on integrated and sustainable crop nutrient practices, is required to obtain high yields from quality planting material. In some cases, the problem of unsatisfactory field performance was traced to nutritional imbalance and poor soil conditions. The commercial micropropagation companies should encourage the availability and use of site-specific agronomic practices by the growers to ensure sustainable success. This would provide farmers the latest technology product with minimal risk. Thus, to achieve expansion of commercial micropropagation, the use of cost-effective technology and of locally adapted varieties and relevant agronomic practices is necessary.

Plant tissue culture refers to growing and multiplication of cells, tissues and organs of plants on defined solid or liquid media under aseptic and controlled environment. Micropropagation allows rapid production of high quality, disease-free and uniform planting material. The micropropagation of high quality planting material of ornamentals, and forest and fruit trees has created new opportunities in global trading for producers, farmers, and nursery owners, and for rural employment. The plants can be multiplied under a controlled environment anywhere irrespective of the season and weather on a year-round basis. However, micropropagation technology is more expensive than the conventional method of plant propagation. It is a capital-intensive industry, and in some cases the unit cost per plant becomes unaffordable. Hence, it is necessary to adopt strategies to reduce production cost and lower the cost per plant.

Plant micropropagation is primarily based on rapid proliferation of tiny stem cuttings, axillary buds, and to a limited extent of somatic embryos, cell clumps in suspension cultures and bioreactors. The cultured cells and tissue can take several pathways. The pathways that lead to the production of true-to-type plants in large numbers are the preferred ones for commercial multiplication. The process of micro-propagation is usually divided into several stages i.e., pre-propagation, initiation of explants, subculture of explants for proliferation, shooting and rooting, and hardening. These stages are universally applicable in large-scale multiplication of plants. The delivery of hardened small micropropagated plants to growers and market also requires extra care.

Low-cost tissue culture technology is the adoption of practices and use of equipment to reduce the unit cost of micropropagule and plant production. Low cost options should lower the cost of production without compromising the quality of the micropropagules and plants. In low cost technology cost reduction is achieved by improving process efficiency and better utilization of resources. Low-cost tissue-culture technology is high priority in agriculture, horticulture, forestry, and floriculture of many developing countries for the production of suitably priced high quality planting material.

A number of low-cost alternatives can be used to simplify various operations and reduce the costs in a tissue culture facility. The physical components of a typical plant tissue culture facility include equipment and buildings with preparation room, transfer room, culture or growth room, hardening and weaning area, soil-growing area (greenhouses, plastic tunnels), packaging and shipping area, and related facilities such as an office, and a store for chemicals, containers and supplies. The size of the physical components of a tissue culture facility will vary according to its functional needs, i.e. the volume of production. Careful planning of a facility can make large savings both in the construction costs and day-to day operations in the facility. It is recommended that an existing facility should be visited to view the layout and operational needs before starting a new facility.

Proper choice of media and containers can reduce the cost of micropropagation. The composition of culture media used for proliferation has a tremendous influence on production costs. The type of culture vessel influences the efficiency of transfer during subculture and production of propagules per unit area. The replacement of expensive imported vessels with reusable glass jars and lids, alternatives to gelling agents, use of household sucrose, and some medium components can reduce costs of production. Bulk making of media and storage as deep frozen stocks also reduces labour costs.

Artificial lighting of cultures in the growth rooms is one of the most expensive and inefficient methods in tissue culture technology. Changing the method of illumination from artificial to natural light is a decisive low cost option in tissue culture. This reduces electricity and capital costs and also improves the plant quality. Maintaining in vitro cultures at a regulated temperature with air conditioners adds to the cost but does not contribute to specific plant quality.

Many in vitro growing plants can tolerate wide fluctuations in temperature, and adapt better to field conditions than those grown under even temperature. Plants can also be hardened in open shade. Plants hardened under natural light are sturdy, and withstand transplantation better in the field. Production of plants based on tissue culture technology and their subsequent growing is a labour intensive system. Even in developing countries, where labour is relatively less costly, hiring expertise from established R&D laboratories reduces overhead costs.

Bioreactor-based propagation of plants can increase rate of multiplication and growth of cultures and reduce space, energy and labour requirements in commercial micropropagation. They can therefore be attractive to developing countries as regards new or expanding plant culture facilities, in combination with a conventional laboratory. Bioreactors provide more precise control of the plant growth gaseous exchange, illumination, medium agitation, temperature and pH than the conventional culture vessels.

However, to be cost-effective, use of bioreactors requires indexed plant cultures, and attention to aseptic procedures during handling of plant material otherwise culture contamination leads to massive economic loss. In micropropagation, the health status of the donor mother plant and of the plants multiplied from it are among the most critical factors, which determine the success of a tissue culture operation.

The indexing of the mother plants for freedom from viral, bacterial, and fungal diseases is a normal procedure in large-scale plant propagation through tissue culture. Plants not originating from pathogen-tested material must be screened for the presence of viruses. Laboratories, which do not have in-house facilities to carry out plant indexing, should obtain their indexed stock plants from organizations such as Departments of Agriculture, agricultural universities or privately owned certified germplasm repositories that routinely produce such plant material.

Batches of micropropagated plants should be tested for freedom from diseases either in-house or by other laboratories. ELISA has been the most effective method for virus and pathogen detection in plants. Polymerase chain reaction and nucleic acid hybridization are more sensitive than ELISA, and can detect pathogens in extremely low amounts. Quality control is essential to assure production of high quality plants and to have end-users confidence. Quality standards require the establishment of suitable tests to maintain quality control.

The choice of explant source, freedom of the donor plant from viruses, disease causing fungi, bacteria, viroids, phytoplasmas, vigour and conformity of the variety, and elimination of somaclones are critical for maintaining plant quality. Variety identification by proper labeling at all stages is essential to ensure varietal identity. Tissue-cultured propagules are produced under a controlled environment. Such plantlets have small juvenile leaves with reduced photosynthetic capacity, and malfunctioning stomata. Priming for rooting, shooting, and improved photosynthesis can be achieved with growth regulators and adjustment to the growing conditions that affect the post-transplanting performance of the propagules.

Vented closures with microbial filters facilitate gas exchange, reduce ethylene build-up that stunts plant growth, reduces leaf-size, and causes leaf drop in tissue culture containers. Plantlets produced under photo-autotrophic culture systems on media with or without sucrose but CO_2-enrichment, increased light intensity, good gas exchange and reduced humidity are more vigorous, have larger root-systems, and are less susceptible to microbial contamination.

Plants adapted gradually to the ex-vitro environment have improved survival upon transfer to soil. Plants in their natural environment are colonized with many bacteria, fungi, and mycorrhizae. In-vitro or ex-vitro biopriming of micropropagated plants with such organisms improves plant performance under stress environments, and consequently enhances yield.

MICROPROPAGATION OF DIEFFENBACHIA

DIEFFENBACHIA

Dieffenbachias are very popular and hardy shade loving plants, commonly grown in pots for indoor decoration and also in shade and semi-shade in gardens of warm humid regions. It has the common name Dumb Cane because of the poisonous nature of their sap, which causes intense pain and swelling of the tongue, if it is taken internally. The leaves of this fleshy plant are oval-oblong, gradually tapering towards the apex, arising from the central stem and spotted, striped or speckled with white or yellow, fifteen to forty cm or so in length.

They grow in any type of well drained soil but organic matter certainly helps to make the leaves more attractive. The plant can be propagated by suckers from the base, by top cuttings or by planting of stem node pieces with a good bud on it. A large number of plantlets are also produced through micro propagation. This plant has got a natural tendency to be lanky and top heavy. In that case cut the stem from where you have new buds or eyes. New shoots will start from that point. Do not discard the top part, put it in a potting mix and keep it moist. new root will start after a few days, and will get one more plant. This plant is not disease prone. Sometimes they can be infested by mites. Treat them as you do for other plants.

DIEFFENBACHIA AMOENA CALCUTTA

It is a sport of D. Amoena, large and thick stemmed. Leaves large, arching, lamina predominantly yellow-white with irregular green spot, margin branded green.

TROPIC SNOW

The scientific name is Dieffenbachia Amoena. It has compact growth, leaves leathery, glossy deep green, highly variegated form, centre with cream and green, border deep green.

CAMILLA

It's scientific name is Dieffenbachia Exotica. It has compact growth, freely branched. Its leaves are small, ovate, pointed apex, predominantly creamy yellow, border green.

DIEFFENBACHIA MACULATA

It is a fairly large hardy plant with glossy green leaves, oval with chordate base, blotched or marbled with white pale green.

DIEFFENBACHIA JUPITER

An outstanding and attractive variety, came from Dieffenbachia Maculata It has many ovate leaves and leaf size is smaller compared to others. Young leaves are yellowish cream and border light green.

MICROPROPAGATION OF LILAC

TISSUE CULTURE OF LILACS

Traditionally lilacs have been grown from seed, suckers or by grafting superior clones on to compatible rootstocks. Lilac seedlings of several species, privet stem cuttings as well as ash seedlings/root cuttings have all been utilized as rootstocks. In recent times there has been a move away from using rootstocks for various reasons, and nurserymen have been trying to produce cultivars on their own roots. While cuttings taken at the right time are often successful for many species, this is not always a sure recipe for thevulgaris clones.

For more than two decades some of the leading nurseries in the US and Europe have been using tissue culture for propagating lilacs on their own roots. While the main reason for resorting to micropropagation is to produce large numbers of selected clones within a short period, we have discovered it to be no less effective in producing small numbers ofvulgaris cultivars which are recalcitrant to initiating roots by conventional means. The following is a short photo-essay on the process.

The most difficult part of the operation is obtaining material free of bacterial and fungal contamination. Succulent terminal cuttings are divided into single node 'explants' and agitated in dilute chlorine for varying lengths of time to try eliminate pathogens.

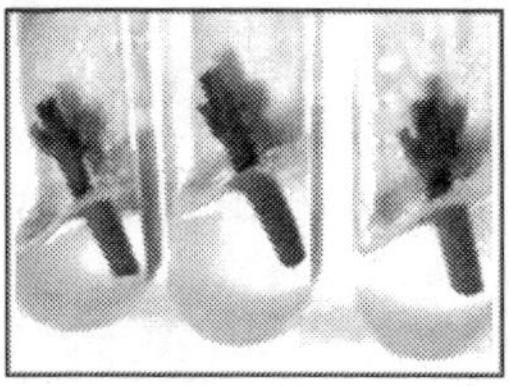

Cleaned explants are placed in nutrient agar under aseptic conditions and stored under lights at a constant fairly high temperature for several weeks.

Explants free of contamination make 20-50 mm of growth over several weeks and are ready for weaning and rooting, or for further division to repeat the process until the required quantity is achieved. Leaves at this stage are tiny, resembling those of young seedlings than that of mature plants.

A tray of cuttings after root initiation and several months' growth in the greenhouse, around 150 mm height.

Our first micropropagated lilac, standing all of 125 mm after a season's growth in the greenhouse. Note the sucker emerging from the left of main stem. A second season's growth should see this on the market.

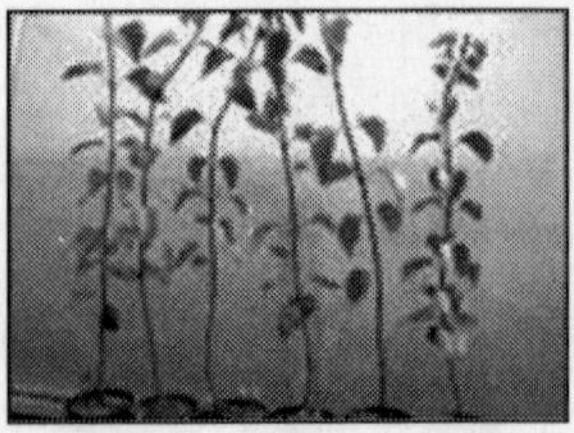

Well grown tissue cultured *S. v.* 'Sensation' towards the end of the second season at leaf-fall, standing nearly 1 m high. Some cultivars such as*S. x hyacinthiflora* 'Maiden's Blush' branch more easily and make less height.

4

Mitochondria: Power Plant of Cell

For thousands of years, men puzzled over the question of where body heat comes from. This heat was recognized as nearly synonymous with life itself–but what produced it? The source of the energy of man or beast was equally mysterious. In recent decades it has become clear that both heat and energy can be traced to some sausage-shaped organelles in our cells, the mitochondria. Of course the ultimate source of energy for all plants and animals is sunlight. But the sun's energy can be harnessed by plants, through photosynthesis and stored in molecules of carbohydrates. When animals eat these carbohydrates and break them down to carbon dioxide and water, with the help of oxygen and an arsenal of enzymes, large amounts of energy become available. Animals immediately convert this energy into molecules of high-energy ATP (adenosine triphosphate) -- the universal currency of energy in living things. Excluding only the very first stages in carbohydrate breakdown, which are called glycolysis, the entire, complicated process of energy transfer to ATP takes place within the mitochondria.

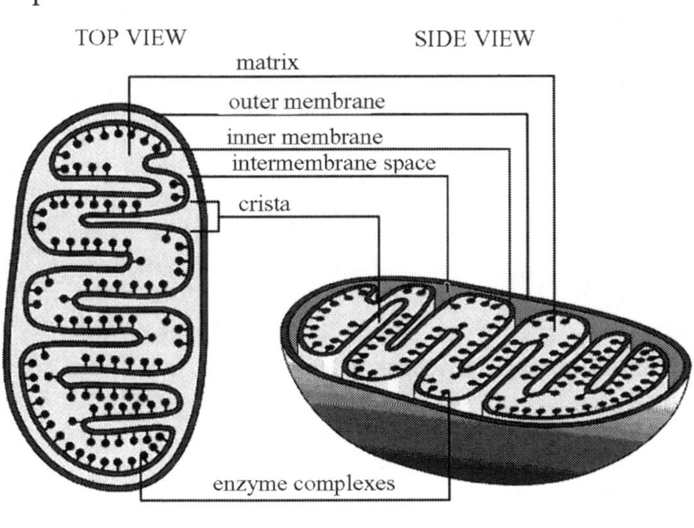

Fig. Mitochondrion.

Mitochondria are the largest organelles in the cell, after the nucleus, yet some cells have more than a thousand of them. Thin and long, they vary in diameter from 0.5 to 1 micrometer and in length up to 7 micrometers. Their thread-like outlines can be seen with a good light microscope.

Although they were first observed in 1850, it took a century to visualize their internal structure and understand their function. Actually the story goes back even further, to the 18th century, when Lavoisier showed that animals require oxygen for respiration. This led chemists to study cellular respiration, the process by which living cells use oxygen to release the chemical energy stored in foodstuffs.

Early in the 20th century biochemists discovered that these reactions fall into two major groups:

1 The carbon pathway, where a series of chemical reactions, each requiring a specific enzyme, breaks down the carbohydrates into carbon dioxide and hydrogen;
2 The hydrogen pathway, which transfers the hydrogen to oxyen in stages, forming water and releasing energy.

The whole process is organized much like an assembly line. In the hydrogen pathway, the hydrogen's electrons pass through an "electron transport chain" made up of enzymes that act as carriers and as the electrons move from enzyme to enzyme, they give up part of their energy, which is stored in molecules of ATP. The ironcontaining enzymes, cytochrome a, b and c, carry out the final stages of the process. All in all, three molecules of ATP are formed for every atom of oxygen that is used up in respiration. For a long time, biochemists studied these reactions in cell extracts without worrying about what parts of the cells might be involved.

One scientist, B.F. Kingsbury of Cornell University, did suggest that mitochondria might be the site of cellular respiration back in 1912, but this was ignored. In the 1940's, using an ultracentrifuge, Albert Claude and his associates isolated various cell particles including mitochondria. Shortly afterwards, in 1948, Albert Lehninger of Johns Hopkins and Eugene Kennedy of Harvard showed that the reactions leading to the synthesis of ATP occurred in the mitochondria. In the early 1950's, Palade and the Swedish scientist Fritiof Sjostrand reported that mitochondria are bounded by a membrane and that they have a system of parallel, regularly spaced inner ridges which were named cristae.

It is now clear that there are, in fact, two membranes around the mitochondria of plants, protozoa, molluscs, insects and man: an outer membrane, set off by a fluidfilled gap; and an inner membrane which is folded inward at various points to increase its surface, forming the cristae, so that the enzyme molecules of the electron transport chain, which are attached to this inner surface in specific sequences, can be packed more densely side by side in the mitochondria.

This general design for mitochondria seems to have existed unchanged, through the evolution of plants and animals, for more than 1.2 billion years. Though mitochondria are generally visualized as ovalshaped, they can change shape quite drastically. They swell and contract in response to various hormones and drugs. Even in very low concentrations, Lehninger found, the thyroid hormone thyroxin will make mitochondria swell and will simultaneously "uncouple" the electron transport chain from the synthesis of ATP, so that even though carbohydrates may be metabolized and oxygen consumed, no ATP will be manufactured.

ATP, on the other hand, makes mitochondria contract. This swelling and contraction appear related to the movement of water in, out and through cells, which is of particular importance in the kidney. The more scientists learn about mitochondria, however, the more their curiosity is aroused, for these are without a doubt the strangest of the cell's organelles. To begin with, mitochondria contain their own hereditary material (DNA), which resembles that of bacteria far more than that of animal cell nuclei. Because of this similarity, many scientists believe that mitochondria are derived from bacteria that infected a primitive cell and evolved into a useful symbiotic relationship with it. As Lewis Thomas puts it, "A good case can be made for our nonexistence as entities.

We are shared, rented, occupied. At the interior of our cells, driving them, providing the oxidative energy that sends us out for the improvement of each shining day, are the mitochondria and in a strict sense they are not ours. They turn out to be little separate creatures, the colonial posterity of migrant procaryotes, probably primitive bacteria that swam into ancestral precursors of our eucaryotic cells and stayed there. Ever since, they have maintained themselves and their ways, replicating in their own fashion, privately, with their own DNA and RNA quite different from ours. They are as much symbionts as the rhizobial bacteria in the roots of beans. Without them, we would not move a muscle, drum a finger, think a thought. My mitochondria comprise a very large proportion of me.

Looked at in this way, I could be taken for very large, motile colony of respiring bacteria, operating a complex system of nuclei, microtubules and neurons for the pleasure and sustenance of their families and running, at the moment, a typewriter, continues Thomas somewhat facetiously.I am obliged to do a great deal of essential work for my mitochondria. My nuclei code out the outer membranes of each and a good many of the enzymes attached to the cristae must be synthesized by me. Each of them, by all accounts, makes only enough of its own materials to get along on and the rest must come from me. And I am the one who has to do the worrying. During the past few years, scientists have begun to learn which proteins and enzymes of the mitochondria are synthesized according to directions in the mitochondrial DNA and which are controlled by the genes in the nucleus. This information would be of considerable importance for the design of antibiotics and drugs, as well as for the understanding of genetic diseases. For example, researchers have found that chloramphenicol–a powerful antibiotic which can cause a fatal anemia–interferes with the synthesis of mitochondrial DNA, while other antibiotics do not. Mitochondria are just as important to the life of each cell as the heart is to the human body.

At least 95 percent of our cells' energy comes from mitochondria. (Only the mammalian red blood cells, which have no nuclei, get along without mitochondria; they obtain whatever energy they need from the partial breakdown of glucose in their cytoplasm). Growing cells continuously synthesize nucleic acids, proteins and other components, for which they need the energy that is stored in molecules of ATP.

Muscle cells which carry out mechanical work, plant cells which pump water against the force of gravity and many other cells also require a continuous source of energy. Besides supplying this energy, mitochondria apparently help to control the concentration of water, calcium and other ions in the cytoplasm.

They are also deeply involved in the breakdown and recycling of sugars, fatty acids and amino acids to recover their energy, as well as in the formation of urea as a waste product, to be excreted through the kidney. These strange organelles may thus hold clues to a variety of seemingly unrelated diseases. They have been linked to some liver diseases (*e.g.*, viral hepatitis, obstructive jaundice and cirrhosis of the liver); some muscle diseases (*e.g.*, familial myotrophic dystrophy); and some kidney diseases.

In all these conditions, pathologists have seen large amounts of materials pile up inside the mitochondria as "inclusions" of varying shapes and sizes, presumably waiting to be processed. What these inclusions consist of remains a mystery, but if scientists could find out, they might discover which specific components (for instance, enzymes) are involved in each disease. In addition, several calcium disorders may be related to mitochondrial defects. For example, abnormal concentrations of calcium (possibly caused by malfunctioning mitochondria) may lead to arthritis, bursitis, or arterial disease.

While all these leads are promising, the greatest spinoff from research on mitochondria may be in the area of energy–not only for the cell, but for general use by mankind. As the oil crisis has made us realise, we presently depend on a limited supply of fossil fuels–coal and oil which are of biological origin but cannot be replenished.

On the other hand, nature contains a far greater resource: a molecular know-how and technology gained over millennia, by which biological systems manufacture ATP from solar energy. This technology awaits us. It exists in very similar, if not identical forms in mitochondria and in their mirror image, the chloroplasts, which are found in green plants, where they do the work of photosynthesis, converting the sun's energy into carbohydrates.

Like mitochondria, chloroplasts have an outer and an inner membrane. They have some of their own DNA. They, too, may have originated as some kind of infection, probably with blue-green algae. And it is on chloroplasts that we ultimately depend for nearly all our energy–that of our bodies as well as that which is stored in deposits of oil and coal. If scientists ever unravel the details of the process by which chloroplasts store solar energy in carbohydrates and mitochondria release it, we may be able to produce unlimited quantities of usable energy.

THE MITOCHONDRIA

Mitochondria have been known for many years. They were first discovered and described by Altmann in 1886 and were put more or less definitely on the cytological map by Benda in 1903. They can be easily demonstrated with suitable dyes. Altmann's aniline fuchsine-picric acid technique shows them up very well. Regaud's iron-hematoxylin method is also very good.

The mitochondria can even be seen in the living cell by staining them intravitally with Janus green and, of course, they show up extremely well with the phase-contrast and interference microscopes. There is not much difficulty, then, in establishing their existence and nature. Their dimensions vary, of course, but they range in most cells from about 0.5 to 2ì or longer. The filamentous mitochondria, for example, which are present in connective tissue cells may be much longer than this. Extremely small and even submicroscopic mitochondria may also exist.

Extremely smallsized mitochondria are, in fact, known and Rhodin has described structures ranging from 0.1 to 0.5ì which are presumably mitochondrial in nature. Green has pointed out that the various enzymes and proteins which mitochondria contain can be accommodated in a very much smaller particle than any known mitochondria and still perform all the functions required of them.

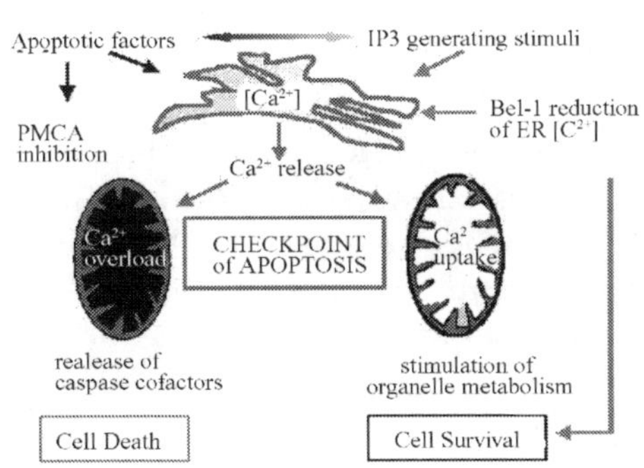

Fig. Mitochondria Ganglion Neurons.

Some of these much smaller mitochondria may not have all the details of organization of larger mitochondria but there is no reason why they should not contain all the enzymes and other compounds which are important for mitochondrial structure and function. It is of interest to note that, in living tissue culture cells, the mitochondria undergo continuous movement.This has been noted by quite a number of workers and the movements have been described frequently.

These movements may be fitful individual movements by the mitochondrion itself, *i.e.*, bending, wriggling and writhing, or may consist of movement of the whole mitochondrion through the cytoplasm of the cell. The cause of the bending and twisting movements of the mitochondria is not exactly known. At one time it was thought to result from the contracting of the polypeptide chains which are part of the membrane of the mitochondria. It may be due to the fact that the mitochondria were engaged in active ion pumping or being twisted by cytoplasmic movements. In the second type of movement in which there is a transport of the whole mitochondrion from one part of the cell to another, the movement is possibly related to streaming movements of the cytoplasm or to electrical forces.

The mitochondria have been described as making journeys from the cell membrane to the nuclear membrane and back again almost as if they were discharging either an electric charge or even perhaps some compound on the membrane that they touch. There are also records that, when the mitochondria come in contact with the nuclear membrane the nucleolus sometimes moves across the cell and comes in contact with the nuclear membrane at the same point at which the mitochondrion is touching it and this suggests that there may be some exchange of compounds or substances during this period. This has been demonstrated in spinal ganglion cells by Dr. Tewari in the author's labouratory.

Filamentous mitochondria have frequently been observed to break up into batonettes and these have been observed to break up into granules. The reverse process has also been found to occur, granules have been seen to join up into batonettes and these into longer filamentous mitochondria. What the significance of this breaking up is we do not know, but a little later we will refer to this process again in the light of what will be said about the metabolic activities of mitochondria.

When mitochondria were studied in ultrathin sections of tissue, it was found by Palade in 1953, Sjöstrand, 1953 and by Rhodin in 1954, that they had an interesting internal structure. These workers found first of all that there was a single membrane round the outside, that inside this membrane was another membrane so that the two made a double membrane and, after a little controversy, it was agreed that the inner of these two membranes was extended to form a number of bars or plates which projected into the interior of the mitochondria, in some cases touching or almost touching the other side. These plates or bars or tubes had double structure again, since they were a reflection of the inner membrane of the mitochondria.

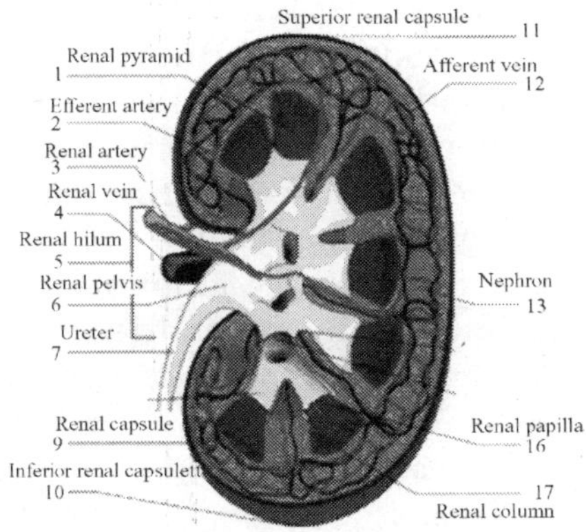

Fig. Mitochondria in Kidney

They were described by Palade as the "cristae mitochondriales". Changes in these cristae appear to be related to function, for instance Palade has drawn attention to the fact that the amount of cytochrome in the mitochondrial fraction of a homogenate is directly related to the number of cristae present in the mitochondria. When metabolic activity is high, for example, in rapidly contracting skeletal and heart muscle, the mitochondria have many densely packed cristae. In smooth muscle, however, where the activity is greatly reduced, the cristae present in the mitochondria are relatively sparse.There are claims that, *e.g.*, in mouse tumor cells and in paramecium, cristae may be everted into the cytoplasm. Presumably in this case the outer membrane is folded into cristae. It is of interest that the double membranes which surround the mitochondria show the same 80 A unit structure which is characteristic of the cell membrane and Robertson has suggested that it is possible that the mitochondria may have originated by an invagination of the membrane which was nipped off to form the mitochondrion.

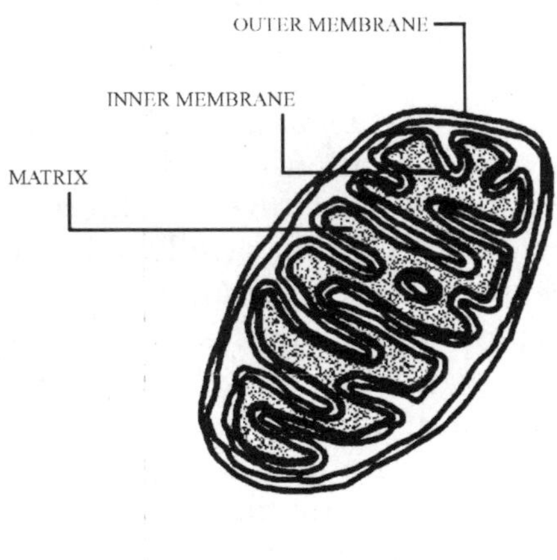

Fig. Structure of Mitochondria

This is an interesting suggestion but it is not readily acceptable to most cytologists and is, in fact, considered unlikely by many of them. The membranes of the mitochondria appear to consist of two protein layers separated by a double layer of lipids. This is again characteristic of cell membrane structure. Bensley in 1934 was the first one to isolate mitochondria by differential centrifugation and he was able to demonstrate that mitochondria contained a considerable amount of protein, lipid and fat. Bensley's chemical composition of mitochondria is

Proteins and unknowns	65per cent
Glycerides	29per cent
Lecithin and cephalin	4per cent
Cholesterol	2per cent

Subsequent studies have shown up to 30per cent of phospholipid (lecithin and cephalin). Mitochondria undergo a very considerable change under various conditions. Bensley claimed that mitochondria (as demonstrated by standard staining reactions) were greatly reduced in liver cells of starved animals, but Sheridan in our labouratory and Fawcett have demonstrated by electron-microscope techniques a great increase of mitochondria in starvation.

Sheridan has further demonstrated a still further increase of mitochondria in the liver cells of guinea pigs suffering from scurvy. Mitochondria are also affected in number and form by narcotics, enzyme poisons, old age and so on. Sometimes mitochondria store unusual compounds and sometimes they store compounds which are usual but to an abnormal degree. The double membrane enclosing the mitochondria is of a semipermeable nature and as a result of this mitochondria have been described as osmometers. There is no doubt that molecule of various sizes pass with different degrees of readiness through the mitochondrial membrane.

It has been shown that large molecules pass through fairly easily while some small molecules pass through either not at all or only with very great difficulty. It has even been said by Emmelot and Bos that this permeability, particularly of liver mitochondrial membranes, is affected by thyroxine. Also there are claims by some authors that mitochondrial membranes have pores. Mitochondria may accumulate water and swell. They may do this in starvation and in various pathological conditions.

When this occurs they produce the condition which the pathologists describe as cloudy swelling. Cloudy swelling has been known to occur sometimes in neo-plasms and sometimes after poisoning with various toxins. The mechanics of cloudy swelling are that the mitochondria become very large and lose their cristae, the matrix stains less strongly and sometimes the outer membranes of the mitochondria may disappear and the mitochondria may fuse and so form structures known as chondriospheres. Under some pathological conditions mitochondria may become disrupted, undergo lysis and portions may be ejected from the cell. Mitochondria are capable of abnormal storage. Accumulation of ferritin by mitochondria was demonstrated by Kuff and Dalton in 1957 and they may also accumulate iron pigments or bile pigments. Brachet has described ferritin particles in young erythroblasts where he says they penetrate the mitochondria and then break into smaller iron-containing particles. When the mitochondria eventually burst, the granules become scattered throughout the cytoplasm and take part in the formation of hemoglobin. Silver granules and keratinous materials have been also seen inside mitochondria. They have been found to contain melanin and Graffi in a series of papers from 1939 to 1941 claims that they also contain or store carcinogenetic hydrocarbons.

Sometimes this storage affects the function of the mitochondria and sometimes the accumulation of material is due to malfunction of the organelle. They may also store neutral fats or lipids. In plants mitochondria have been recorded as storing starch. As a result of storage mitochondria undergo physical changes. The membrane may become single and the cristae decrease in size and may be lost altogether. If the stored substance is eliminated, the mitochondria resume their normal appearance. There is a vast litreature on the role of mitochondria in the production of almost any kind of product—protein, yolk, fat, glycogen and so on—but this is not the place to review this work. There have been many scornful remarks in the litreature about the supposed synthetic activity of mitochondria, but the study of the enzymic equipment of mitochondria shows that they have the equipment to effect a wide range of synthetic activities.

It has been mentioned that the isolation of mitochondria by the differential centrifugation of homogenized cells was first carried out in 1935 by Bensley and by Hoerr. It was not until 10 or 15 years later that this technique became used extensively and this time by biochemists who set to work to find out the metabolic composition of mitochondria, *i.e.*, the nature and content of enzymes concerned with metabolic processes.

STRUCTURE OF MITOCHONDRIA

A mitochondrion consists of two membranes and two compartments, the larger of which contains the *mitochondrial matrix*. An outer limiting membrane, about 6 nm thick, surrounds the mitochondrion. Within this membrane, and separated from it by a space of about 6 to 8 nm, is an inner membrane that projects into the mitochondrial cavity complex infoldings called *mitochondrial crests*.

This inner membrane, also about 6 nm thick, divides the mitochondrion into two chambers or compartments:

1. The outer chamber contained between the two membranes and in the core of the crests
2. The inner chamber, bound by the inner membrane.

This inner chamber is filled with a relatively dense proteinaceous material usually called the *mitochondrial matrix*. This is generally homogeneous, but in some cases it may contain a finely filamentous material, or small, highly dense granules. These granules are now considered to be sites for binding divalent cations, particularly Mg^{2+} and Ca^{2+}. The mitochondrial crests that project from the inner membrane are, in general, incomplete septa or ridges that do not interrupt the continuity of the inner chamber; thus, the matrix is continuous within the mitochondrion.

THE MITOCHONDRIAL MATRIX CONTAINS RIBOSOMES AND A CIRCULAR DNA

Within the mitochondrial matrix are e-mail ribosomes and a circular DNA. The outer and inner membranes and the crests can be considered to be fluid molecular films with a compact molecular structure, the matrix is gel-like and contains a high concentration of soluble proteins and smaller molecules. This double (solid-liquid) structure is important in providing an explanation for some of the mechanical properties of mitochondria.

F_1 Particles

The use of negative staining has enabled recognition of other details of mitochondrial structure. If a mitochondrion is allowed to swell and break in a hypotonic solution and is then immersed in phosphotungstate, the inner membrane in the crest appears covered by particles of 8.5 nm that have a stem, linking each with the membrane. These so-called "elementary" or "F_1" particles, are regularly spaced at intervals of 10 nm on the inner surface of these membranes.

According to some estimates, there are between 10^4 and 10^5 elementary particles per mitochondrion. These particles correspond to a special ATPase involved in the coupling of oxidation arid phosphorylation. The presence of such F_1 particles on the matrix side (M side) confers to the inner mitochondrial membrane a characteristic asymmetry which is of basic importance to its function. It is interesting to recall here that bacterial plasma membranes contain F_1 particles with a similar asymmetric distribution (*i.e.*, towards the bacterial matrix). On the other hand, the opposite orientation is found in chloroplasts where there are F_1 particles on the outer or stromal side of the thylakoid membrane.

Different Cell and Mitochondrial Sturctural Variation

Some investigators assume that a common pattern of mitochondrial structure developed at an early stage of evolution and was subsequently transmitted, without considerable modifications, from protozoa to mammals and from algae to flowering plants. Detailed structural variations can be observed, however. The crests may be arranged longitudinally or they may be simple or branched, forming complex networks. In protozoa, insects, and adrenal cells of the glomerular zone, the infoldings may be tubular instead of lamellar and the tubules may be packed in a regular way. The number of crests per unit volume of a mitochondrion is also variable. Mitochondria in liver and germinal cells have few crests and an abundant matrix, whereas those in certain muscle cells have numerous crests and little matrix. In some cases the crests are so numerous that they may have a quasi-crystalline disposition. The greatest concentration of crests is found in the flight muscle of insects. In general, there seems to be a correlation between the number of crests and the oxidative activity of the mitochondrion.

On majore feature of the mitochondria is the enlargement of the space within the crests or tubules which, corresponds to the outer compartment of the mitochondrion. This is very conspicuous and appears as tubular or vesicular openings, which are much less opaque than the inner matrix. These structures seem to be related to the specific activity of the gland. In this case, mitochondria, in addition to functioning in cell oxidations, are actively engaged, in the synthesis of steroid hormones.

MITOCHONDRIA AND THE LIPID

Various authors since the time of Altmann have observed that the *disposition of lipids* may be related to mitochondrial activity. In pancreas and liver cells, for instance, after a short period of starvation, the mitochondria come into contact with lipid droplets. The relationship may be so tight that only the inner mitochondrial membrane can be seen adjacent to the lipid in some regions. The electron microscopic images suggest that an active process of fat utilization takes place under the action of the fatty acid oxidases present in mitochondria.

Intramitochondrial Inclusions

An accumulation of pigment derived from hemoglobin has been observed in mitochondria of amphibians. Ferritin molecules accumulate within mitochondria in subjects suffering from Cooley's hereditary anemia.

Another example is the transformation of mitochondria into yolk bodies in eggs of the mollusks. In these, the masses of protein molecules may assume a regular crystalline disposition.

Mitochondrial Sensitivity to Cell Injury and Degeneration

In situation of injury to the cell mitochondria is provide sensitive indicator. Although mitochondria may be readily altered by the action of various agents, the changes are, within certain limits, reversible. If the alteration reaches a certain critical point, however, it becomes irreversible, and this is generally considered degeneration of the mitochondria.

Essentially there are three types of change:

1. Fragmentation into granules, followed by lysis and dispersion,
2. Intense swelling with transformation into large vacuoles,

3. A great accumulation of materials with transformation of mitochondria into hyalin granules. This last change is characteristic of the so-called cloudy swelling and hyalin degeneration that frequently results in cellular death.

A relatively frequent observation in an otherwise normal cell is the presence of degenerating mitochondria in foci of autolysis, constituting a type of lysosome (cytolysosome). Another type of degeneration is the fusion of mitochondria to form large bodies called *chondriospheres.* This degeneration has been found in patients with scurvy and seems to be normal in the adrenal gland of the hamster.

ISOLATION OF MITOCHONDRIAL MEMBRANES

Morphologically homogenous fractions of mitochondria may be isolated from liver, skeletal muscle, heart, and other tissues. Since the early investigations of Hogeboom and others it has been demonstrated that the mitochondrion has a lipoprotein compositon—65 to 70 per cent protein, and 25 to 30 per cent lipid. Most of the lipid content consists of phosphatides (*e.g.,* lecithin and cephalin); cholesterol and other lipids are present in small amounts. Ribonucleic acid was consistently found in about 0.5 per cent of the dry weight.

In recent years the two mitochondrial membranes and the compartments they limit have been separated by density gradient centrifugation. The outer membrane can be separated by causing a swelling, with breakage followed by a contraction of the inner membrane and matrix. The outer membrane is much lighter and needs stronger centrifugal forces or a less dense gradient to be separated.

Mitoplast has been produced by separating these two elements from the outer membrane with digitonin. The mitoplast has pseudopodic processes and is able to carry out oxidative phosphorylation. This separation has provided a clearer definition of the exact localization of the mitochondrial enzyme systems and has revealed interesting differences between the two membranes.

Structural and Chemical Changes in Outer and Inner Membranes

The outer membrane fraction has a 40 per cent lipid content (compared to 20 per cent in the inner membrane), contains more cholesterol, and is higher in phosphatidyl inositol; on the other hand, it is lower in cardiolipin. A fundamental difference between the outer and inner membrane is in the lipid/protein ratio (*i.e.,* about 0.8 in the outer membrane and about 0.3 in the inner membrane).

This low lipid/protein ratio indicates that in the inner membrane there is a greater degree of intercalation of the protein within the lipid bilayer. This fact of structure has been confirmed by electron microscopic observations with freeze-fracturing methods. From the morphologic viewpoint the outer membrane lacks the elementary particles that, are prominent in the inner membrane.

Mitochondrial Enzymes are Highly Compartmentalized

The *outer membrane of mitochondria* contains an NADH-cytochrome-c-reductase system which consists of a flavoprotein and cytochrome b_5. The most specific enzyme system of the outer membrane is *monoamine oxidase,* which may serve as an enzyme marker. This membrane also contains kynurenine hydroxylase, fatty acid co-enzyme A ligase, a phospholipase, and various enzymes of the phospholipid metabolism. The supernatant, obtained after treatment with digitonin is considered to originate from the outer compartment between the two membranes. In this compartment are found adenylate kinase, nucleoside diphospho-kinase, DNAase I, and 5'-endonuclease.

The *mitoplast fraction* (inner membrane plus matrix) contains the components of the respiratory chain and oxidative phosphorylation and the soluble enzymes present in the matrix. The *inner membrane* carries all the components of the respiratory chain and the oxidative phosphorylation system (see later discussion).

It also has glycerol-phosphate dehydrogenase, choline-dehydrogenase, and several carriers *or translocators* for the permeation of phosphate, glutamate, aspartate, ADP, and ATP. It is calculated that the respiratory chain represents about 20 per cent and the phosphorylating system about 15 per cent of the total protein content of the membrane. These impressive figures give an idea of the importance of these systems in the inner membrane and mitochondrial crests.

The inner membrane contains a remarkably high concentration of cardiolipin (polyglycerophosphatides), which appear to be important in all systems involving electron transport. The *inner compartment* or *mitochondrial matrix* contains all the soluble enzymes of the citric acid or Krebs cycle and those involved in the oxidation of fatty acids. The matrix also contains DNA, ribosomes, and other RNA species and enzymes involved in the synthesis of protein (see later discussion).

The Inner Membrane

It has been proved by various studied that the inner mitochondrial membrane has two major portions, one corresponding to the *mitochondrial crests* and the other to the region that lines the *outer membrane.* For example, it has been reported that the F_1, particles and the coupling factors are present only in the cristae. Furthermore, the activities of several enzymes (*i.e.,* succinate dehydrogenase, b-hydroxybutyrate dehydrogenase, rotenonesensitive NADH-cytochrome-c-reductase, and ATPase) predominate in the cristae. Studies in recent years have focused attention on the fact that mitochondrial membranes have a net electronegative charge due to anionic sites (*i.e.*, the membranes move towards the anode in an electrophoretic field).

Using *polycationic ferritin* as an electron microscopic probe for studying the density and distribution of anionic ionic sites, it was found that while the crests have few sites, there are patches of high density in the regions lining the outer membrane. These areas may correspond to those found in the localization of anionic glycoproteins.

Another regional difference of the inner membrane involves its association with the outer membrane at the *contact zones.* These are observed especially when the mitochondria are in the condensed state; there are about 100 such contacts per liver mitochondrion. At each of these points is a concentration of cytoplasmic ribosomes, which has led to the suggestion that the contact zones could be the sites through which cytoplasmic polypeptides enter the mitochondrion.

FUNCTION OF MITOCHONDRIA

The various functions of mitochondria are so intimately related to their molecular structure that one topic cannot be studied without the other. This section requires some knowledge of the enzymatic mechanisms in which mitochondria are involved. To know the complexity of the enzyme system in numerical terms, in a mitochondrion more than 70 enzymes and co-enzymes and numerous cofactors and metals essential to mitochondrial functions work together in an orderly fashion.

Besides oxygen, the only fuel that a mitochondrion needs is phosphate and adenosine diphosphate (ADP); the principal final products are ATP plus CO_2 and H_2O. The three major foodstuffs of the cell (carbohydrate, fat, and protein) are ultimately degraded in the cytoplasm to acetate, a two-carbon unit that is bound to co-enzyme A to form acetyl co-enzyme A. This penetrates the mitochondrion, and the acetate group condenses with *oxaloacetic acid* to form citric acid, a six-carbon compound.

Citric acid is oxidized and loses two carbons as CO_2. In this way the four-carbon compound *succinic acid* is formed, and later on it is oxidized to oxaloacetic acid, starting a new cycle. At each turn of the circle one molecule of acetate penetrates and two CO_2 are released (for more details on the Krebs cycle see textbooks of biochemistry).

An important consequence of the Krebs cycle is that at each turn four pairs of hydrogen atoms are removed from the substrate intermediates by enzymatic dehydrogenation. These hydrogen atoms (or the equivalent pairs of electrons) enter the respiratory chain, being accepted by either NAD^+ or FAD. Three pairs of hydrogens are accepted by NAD^+, reducing it to NADH, and one pair by FAD, reducing it to $FADH_2$ (this last pair comes directly from the succinic dehydrogenase reaction). Since it takes two turns of the cycle to metabolize the two acetate molecules that come by glycolysis from one molecule of glucose, a total of six molecules of NADH and two of $FADH_2$ are formed at the starting points of the respiratory chain.

Electrons Flow Along a Cytochrrome Chain

The early work of Warburg and Keilin using spectrophotometric techniques led to the concept that cell oxidations are brought about by electron carriers arranged in a chain of increasing *oxidation-reduction potential.* Keilin employed a simple hand spectroscope to study insect muscle and found that there were special pigments which underwent changes with oxidation and reduction during respiration. Such pigments, which had special absorption bands in the reduced state, were called *cytochromes* and were later on demonstrated to be *iron-containing proteins.*

In each cytochrome the iron may be in either a reduced (Fe^{2+}) or an oxidized state (Fe^{3+}), and the electrons flow along a cytochrome chain in which the redox potentials become more positive. With the exception of *cytochrome c,* which is a peripheral protein that can be extracted with salt solutions, these pigments are tightly bound to the inner mitochondrial membrane (as intrinsic proteins; and detergents are needed to solubilize and extract them. The electron transport chain does not start with the cytochromes but with the co-enzyme *nicotine adenine dinucleotide* (NAD^+), which is reduced to NADH.

The couple NAD^+/NADH has the most negative redox potential (Eo = -0.32 volts) of the chain and the highest free energy (*G* = 52.7 kcal/mol), which is released upon oxidation. The reversible NAD^+/NADH system is followed in the chain by a co-enzyme Q_{10} (ubiquinone), a flavoprotein containing FMN and FAD, and then by the series of cytochromes to end with O_2, which has the most positive redox potential (Eo = 0.82) with no free energy to be utilized (G = 0). This sequence of electron transfer reactions is consistent with the redox potentials of the various electron carriers, which become more positive as electrons pass from the substrate to oxygen.

Furthermore, sensitive spectrophotometric methods developed by Chance have confirmed the functional nature of this sequence, showing that NAD^+ is the most reduced member of chain and that cytochrome a and a_s (which form part of *cytochrome* oxidase) are the most oxidized. There are additional components, and electron transport chain is much complex it has been found in recent years.

In recent years it has found that the electron transport chain is extremely complex and that there are additional components. Of particular interest is *co-enzyme Q (CoQ) or ubiquinone* (because it is ubiquitous in cells), which is a lipidsoluble benzoquinone with a long side-chain of ten isoprenoid units (thus the name CoQ_{10}) It is thought that CoQ is a kind of shuttling system between the flavoproteins and cytochrome b In addition, the mitochondrial respiratory chain contains enzyme-bound copper and iron-sulfur proteins.

The Respiratory Chain

Previously discussed components of the respiratory chain, as well as those involved in the mechanism of phosphorylation, are integrated within the molecular structure of the inner mitochondrial membrane. It is important to remember that this membrane has a high protein content (60 to 70 per cent) and that most of these proteins are of the integral or intrinsic type, which are intercalated into the lipid bilayer. In fact it is calculated that one third of this membrane is occupied by phospholipids and the rest by protein.

As in the case of the plasma membrane, there is here a certain degree of *fluidity* that allows for the lateral diffusion of the proteins. This has been demonstrated by freeze-fracturing of the membrane. The study of the distribution of the intramembrane particles under conditions in which there is a slow freezing of the mitochondria shows that this treatment causes aggregation of the particles into patches, while rapid freezing gives a random distribution.

The respiratory chain, together with the phosphorylating system, represents about 35 per cent of the protein of the inner membrane; the rest is represented by the various translocators and structural proteins. By the using mild detergent we can saperate the component of respriratory chain into mulitmolecular complexs. Such complexes have a considerable amount of lipid which is essential for their activity.

In fact, if the lipids are extracted, the proteins denature or aggregate, and thus become unable to transfer electrons. Extraction of phospholipids by acetone inactivates some catalytic functions of the membrane; however, restoration of function can be achieved by addition of the lipids and CoQ_{10}

Green and associates have recognized four main complexes (I to IV) which, if mixed in correct stoichiometric ratios, can reconstitute to form the electron transport chain. Complex II corresponds to succinate-Q-reductase (*i.e.*, succinate dehydrogenase), which can transfer electrons directly from succinate to CoQ.

Complex I (NADH-Q-reductase)

NADH is the largest complex, with a molecular weight well above 500,000 and a structure consisting of 15 subunits. It contains as prosthetic groups *Flavin mononucleotide* (FMN) arid six iron-sulfur centres. The NADH reaction site is at the M side; contact between this protein and CoQ *is* apparently made in the middle of the membrane, in the hydrophobic area. Complex I spans the inner mitochondrial membrane and is able to translocate protons across it, from the M side to the C side. This is probably operated through a very hydrophobic subunit of the complex.

Complex II (Succinate-Q-reductase)

Reductase complex is composed of two polypeptides with a total molecular weight of 97,000 daltons. It consists of three iron-sulfur centres and *flavin adenine dinucleotide.*

The succinate binding site appears to be on the M side, and the Q site, in the hydrophobic domain of the membrane. In contrast to complex I, succinate-Q-reductase apparently is unable to, translocate protons across the membrane.

Complex III (QH2-cytochrome-c-reductase)

This complex contains a number of subunits with a total molecular weight of about 280,000 daltons. It consists of cytochromes b, cytochrome c_1, and iron-sulfur protein. A proton pump linked to one of the b cytochromes. has been postulated. In the topology of this complex the Q-site may be in the middle of the membrane in the hydrophobic area, and the cytochrome c site, on the C side.

Complex IV (Cytochrome-c-oxidase)

Cytochrome oxidase is a large complex consisting of several polypeptides and having two cytochromes (a and a_3) and two copper atoms. The Molecular weight is about 200,000 daltons. This complex is thought to traverse the mitochondrial membrane, protruding on both surfaces. Such a *transmembranal orientation* is associated with the *vectorial transport of protons* across the membrane. While experimenting with yeast mitochondria it has been shown that cytochrome oxidase is made of seven subunits that can be resolved by gel electrophoresis and that range between 40,000 and 5000 daltons. It is of great biological interest that the three larger polypeptides are very hydrophobic and that their synthesis occurs in mitochondrial ribosomes, and that the four small subunits are relatively hydrophilic and originate in the cytoplasm.

The seven subunits are arranged in the membrane in a functional sequence, being in contact with cytochrome c on the C side. The electrons then pass to cytochrome a (also in the C side), then to Cu^{2+} (possibly in the hydrophobic subunits); and finally to cytochrome a_3 and oxygen at the M. Using antibodies against cytochrome oxidase, it was found that this complex was inhibited when the antibodies were applied to either side of the inner membrane.

Such a finding is in agreement with the above-mentioned traverse topography of complex IV. By the use of freeze-fracturing it has been found that cytochrome oxidase, like other integral proteins, is highly mobile in the lateral plane of the membrane. This lateral diffusion is inhibited by the action of cross-linked antibodies.

Electron Transport

The electron transport system of mitochondria is *coupled* at three points with the phosphorylating system consult. The protons (H^+) originating from electron transfer are translocated by the respiratory chain across the membrane from the M side to the C side. According to the *chemiosmotic hypothesis,* this translocation creates a pH difference and a membrane potential Both constitute the potential. Both constitute the proton-motive force that tends to move H^+ from the C side back to the M side of the membrane is highly impermeable to H^+ ions, these can only reach the M side through the "proton channel" of the ATPase. When H^+ moves from the C side to the M side, the F_1-ATPase, operating in reverse, catalyzes ATP synthesis. Conversely, when the F_1-ATPase hydrolyzes ATP, it functions as a "proton pump" and ejects H^+ from the M side to the C side.

Mitochondrial ATPase

The mitochondrial ATPase molecule is characterised by a multi-polypeptide complex.

Three parts can be classified:

- The *F_1 particle* or soluble ATPase, which is in the head piece of the matrix side of the mitochondrial crests F_1 contains five associated subunits.
- A complex of very hydrophobic proteins (*i.e.*, proteolipids) localized in the membrane bilayer and probably containing the proton translocating mechanism.
- A protein stalk that connects the two. This portion corresponds to the *oligomycin-conferring protein* (OSCP) and to another protein needed to bind F_1 to the membrane.

Through the elegant experiment of Racker the Function of the phosphorglating system can be best demonstrated. If isolated mitochondria are subjected to ultrasonic fragmentation, vesicles are formed that show a reverse organization ("inside-out" vesicles), with the F_1 particles situated on the outer surface and with he C-side facing the interior of the vesicle. These vesicles are able not only to respire but also to phosphorylate. After treatment with urea, which removes the F_1 coupling factor, the membranes no longer phosphorylate. Reconstitution of the complete phosphorylating system can be achieved by mixing the non-phosphorylating membranes with the F_1 particles. After it is isolated, the F_1 particle causes the hydrolysis of ATP to ADP and phosphate; however, when it is attached to the membrane it functions as a synthesizing enzyme.

Respiratory Chains Topological Organization and the Phosphorylation System

All that was just said about the various complexes of the respiratory chain and the phosphorylating system indicates that the inner mitochondrial membrane is highly asymmetrical in its molecular organization. In other words, there is a special sidedness" of the most important components. Under the electron microscope the sideness has been demonstrated by the use of antibodies and also by reagent a that produce electron-dense deposits in the mitochondrial crests.

By using both a special tetrazole, which is converted to formazine by succinate dehydrogenase and 3′,3′-diaminobenzidine (DAB), which is oxidized by cytochrome c, it has been shown that the dense products accumulate on the outside surface of the mitochondrial crest. From experimental evidence it can be demonstrated that the electron transport system is accessible to NADH and succinate only from the inner matrix side, while, cytochrome c is reached from the outer surface of the membrane. In addition to this transverse topology there is also a lateral topology of the components of the respiratory chain and oxidative phosphorylation. The current theory is that the compact assemblies of macromolecules are arranged in a mosaic which is regularly spaced of by other proteins and lipids.

The number of assemblies varies according to the tissue and the folding of the membrane. A mitochondrion from liver has about 15,000 assemblies, while one from the flight muscle of an insect may have as many as 100,000 Within this assembly or *unit of phosphorylation* the main enzymes are in equimolecular quantities. This finding has resulted from the use of ultrarapid spectrophotometric studies and from the determination of binding sites of drugs that specifically affect the various components of the system. For example, *aurovertin* binds to the ATPase; *oligomycin* acts at the level of the binding of ATPase to the membrane; *antimycin* is an inhibitor acting on the QH_2 cytochrome-c-reductase; *rotenone* acts on NADH-Q-reductase; and *cyanide* has its well known poisoning affect on cytochrome-c-oxidase.

The minimum unit of oxidative phosphorylation contains numerous proteins with a total molecular weight approaching 5106 daltons. This unit or assembly must, therefore, cover a considerable area of the membrane. Form biochemical and electron microscopic studies it is possible to imagine a mosaic organization in which each unit os phosphorylation should cover a surface area of about 20 × 20 nm.

The Chemiosmotic Hypothsis

The above-mentioned studies have demonstrated that for each pair of electrons received from the Krebs cycle and transported by the electron transfer system there is one molecule of ATP synthesized by the ATPase. The fundamental problem is that of establishing the nature of the link between these two systems.

The main feature of the *chemiosmotic coupling* proposed by mitchell is that the link is essentially electrochemical in nature. This model is based on the fact that the mitochondrial membrane is impermeable to H^+ and OH^- ions, so that pH differences, acting as energy-rich gradients, can be produced across, it. In this hypothesis the electron transport system is organized in "redox loops" which translocate protons from the M side onto the C side of the membrane. This vectorial movement of protons gives rise to a pH difference and an electrical potential across the membrane.

The chemiosmotic hypothesis postulates that the primary transformations produced by the respiratory chain guide the *osmotic work* needed to accumulate ions. The energy generated by electron transport is conserved in the energy-rich from of a H^+-ion gradient. The pH difference due to the high concentration of OH^- at the M side and of H^+ at the C side also maintain the electrical gradient.

The action of the proton pump of the ATPase, drives the oxidative phosphorylation of ADP to from APT, by which mechanism free energy is conserved:

$$ADP + Pi = ATP + H_2O$$

In this reaction H_2O is also formed because of the dehydration which leads to the removal of H^+ and OH^- ions.

A special feature of this hypothesis is that both the respiratory chain and the ATPase may operate independently. Since coupling by electrical fields can operate at *long-range* distansces there is no need for direct contact between the macromolecules involved. In recent years much experimental evidence has supported the validity of Mitchell's chemosmotic hypothesis, and the mechanism of the redox proton pump has been elucidated by experiments in which various components of the electron transport and phosphorylating system were reconstituted into vesicular bilayers, the so-called liposomes.

The Chemical

An alternate theory—the chemical-conformational hypothesis-postulates that the electron transport and phosphorylating systems are in molecular contact and that information is transmitted by short-range interactions which may, in part, be electrostatic in nature. It is postulated that upon acceptance of the two electrons, each electron transfer protein undergoes a conformational change which is then transmitted to the ATPase by the short-range interactions.

Mitochondria and its Permeability

Metabolic activities of mitochondria take place within the inner mitochondrial compartment, there should be a rapid and active flow of certain metabolites across the two membranes. The products of extramitochondrial metabolism, for example, must reach the mitochondrial matrix in order to undergo oxidation and ADP and phosphate must enter to from ATP. Simultaneously, end products such as H_2O, ATP, urea, and ammonia must leave the mitochondrion.

There are important differences in permeability between the mitochondrial membranes. The outer membrane is freely permeable to electrolytes, water, sucrose, and molecules as large as 10,000 daltons. The inner membrane, is normally impermeable to ions, sucrose.

ADP, ATP, and Pi- Transport by Specific Carriers

The inner mitochondrial membrane uses specific *carriers* or *permeases* for the translocation of various substances.

In mitochondria the existence of specific carriers for ATP (or ADP), phosphate, succinate (or malate), isocitrate, glutamate, aspartate, and bicarbonate has been suggested. A carrier mechanism has been postulated for Ca^{2+} Mn^{2+},or Sr^{2+}.

From above given the dominant carriers are those involved in the *transport of ADP, ATP, and Pi* (inorganic phosphate) through the inner mitochondrial membrane. Such mechanisms provide for the entrance of ADP and Pi into the mitochondrial matrix and for the exit of ATP to the cytoplasm. Study of the ADP-ATP translocation is facilitated by the fact that there are specific inhibitors of this transport, the experimental use of which has pointed to the *gated pore* as the most plausible of the models.

Such an ADP-ATP carrier would be asymmetrically disposed through the membrane, with two conformations, one facing the C side, and the M side. The same carrier could be used for the reverse translocation of ADP and ADP. Isolation of the carrier protein has been achieved by the use of ^{35}S *carboxyatractylate* (^{35}S CAT), which inhibits transport at the C side, followed by the use of detergents. In this way, a ^{35}S CAT-carrier complex having a molecular from mitochondria. Two of these subunits probably constitute the carrier.

Mitochondrial Swelling and Contraction

During swelling the mitochondrial volume may increase three to five times its normal value in the absence of ADP. Mitochondria again reach to its origenal size with the addition of ATP or restoration. Studies on the phenomenon of swelling demonstrate that mitochondria have an important function in the uptake and extrusion of intracellular fluid.

Agents known to induce swelling are phosphate, $Ca^{2+,}$ reduced glutathione, and, in particular, the thyroid hormone thyroxine. Thyroxine is the most effective swelling agent and is capable of producing swelling at physiologic concentrations.

Mitochondrial Conformation

Mitochondria may alter their internal conformation between the two extreme states. One is the so-called *orthodox* state that is usually observed in intact tissues. The other corresponds to the *condensed* state, in which there is a dramatic contraction of the inner compartment of the mitochondria accompanied by accumulation of fluid in the outer compartment. In the orthodox conformation the inner membrane shows the characteristic crests; the matrix fills practically the entire volume of the mitochondrion and has a reticular or granular aspect. In the condensed conformation the inner membrane is folded and the matrix, now more represents only about 5 per cent of the mitochondrial volume. In this state, at certain points, it is possible to see contact zones between the inner and outer membranes of the mitochondrion.

The electron transport system is required for the change from condensed to orthodox conformation to take place. Inhibition of the respiratory chain by cyanide, antimycin A, or Amytal will impair this transformation. The orthodox state is induced when the external ADP becomes low and there is none left to be phosphorylated. If at this time DAP is added, respiration is rapidly enhanced, and the contraction of the inner membranes takes place. It is thought that during the transition from the orthodox to the condensed stage there is change both in the inner membrane and in the matrix. The inner membrane is believed to contain the contractile elements or "mechano-enzymes," but the possible role of the matrix in this contraction should also be considered.

Mitochondrial Accumulation of Ca^{2+} and Phosphate

Respiration and oxidative phosphorylation are the most important functions of mitochondria, another related function is the accumulation of cations. Ca^{2+} can by concentrated in isolated mitochondria up to several hundred times normal values. Phosphate enters the mitochondria along with Ca^{2+}. The amounts of Ca^{2+} and phosphate accumulated may be so great that the dry weight may increase by 25 per cent, and microcrystalline, electron-dense deposits may become visibles within the mitochondria. Electron microscopy coupled with high resolution micro-incineration have been used to study these deposits. This process usually occurs in the osteoplasties present in tissues undergoing calcification.

In the presence of Ca^2, mitochondria no longer phosphorylate but infact accumulate Ca^{2+} and phosphate. Both oxidative phosphorylation and accumulation of Ca^{2+} depend on the maintenance of cell respiration (*i.e.*, the Krebs cycle). The accumulation of Ca^{2+} and other cations inside the mitochondrion is accompanied by loss of H^+, so as to maintain an electrical equilibrium. The matrix becomes more alkaline because of OH^- accumulation, while acid is simultaneously released to the outside. Interestingly enough, chloroplasts behave precisely in the reverse direction by pumping out OH^- and accumulating H^+.

Mitochondria and its Biogenesis

Two main mechanisms for the biogenesis of mitochondria have been postulated: origination by division from parent mitochondria or origination *de novo from* simpler building blocks. Mitochondria are distributed between the daughter cells during mitosis, and their number increases during interphase. It has been observed with time-lapse cinematography that mitochondria gradually elongate and then fragment into smaller mitochondria. This observation has been verified in *Neuropora.*

After labeling of a choline deficient mutant of *Neurospora* with radioactive choline, the radioactivity was followed in the mitochondria of the second and third generations. By radioautography it is found that all mitochondria of the original progeny were labeled. The mitochondria of each daughter cell were also labeled but contained about half the radioactivity.

This seemed to indicate that mitochondria had divided and grown by the addition of new lecithin molecules to the existing mitochondrial framework. Yeast cells grown anaerobically lack a complete respiratory chain (cytochromes b and a are absent); under the electron microscope they show no typical mitochondria; When the yeast cells are placed in air, however, the mitochondrial membranes present fuse, unfold, and from true mitochondria that contain the cytochromes. This observation is somewhat in agreement with the *de novo* synthesis theory of the biogenesis of mitochondria.

The chemical composition of the outer membrane is similar to that of the endoplasmic reticulum and very different from that of the inner membrane Furthermore, continuities between the outer membrane of the mitochondrion and the endoplasmic reticulum, and even with the axon membrane, have also been observed.

THE CHEMICAL NATURE OF MITOCHONDRIA

In Bensley's work it was suspected that mitochondria contained a good deal of lipid material. One finding which suggested this was that mitochondria were stained by a method which is similar to that used for staining *Mycobacterium tuberculosis* and *Mycobacterium leprae*. These bacteria are stained by treating them for some minutes with a hot phenolic solution of basic fuchsine and once stained in this way they resist the de-staining effects of acid alcohol.

This is the origin of their designation as acid-fast bacteria and their staining idiosyncrasies are believed to be due to a waxy or lipoidal coat. Mitochondria stained by hot acid fuchsine resist the decoloring action of picric acid for longer periods than most other cell constituents and this suggests that the mitochondria might also have a membrane which contains material of a lipoidal nature or at any rate which is of similar composition to that of the bacteria mentioned.

Mitochondria are difficult to demonstrate with conventional methods of staining if the fixative has contained acetic acid, alcohol, ether, chloroform, acetone, or other fat solvents. Also mitochondria have been found to stain with osmic acid and, incidentally, to resist de-staining by extraction with turpentine—a property of certain types of lipoprotein complexes. Baker has found that his "acid-haematin" test for lipids applied to a variety of tissues, nearly always gave a positive, reaction with the mitochondria. Thus there had accumulated for some years direct and indirect evidence that mitochondria contained appreciable amounts of lipid. Bensley's experiments, which, incidentally, were carried out some years before Baker's staining studies, showed that mitochondria which had been produced by homogenization and differential centrifugation contained appreciable amounts of lipid (30per cent).

METABOLIC SUBSTANCES FOUND IN MITOCHONDRIA

Perhaps we should give a brief summary of what was suspected about the metabolic significance of these organelles prior to the revolutionary studies which were made with isolated mitochondria beginning during the middle of the 1940's. One of the characteristic staining reactions of these organelles is that they give a green or green-blue stain with Janus green B which is diethylsafranine azodimethylaniline and it has been claimed by Cowdry that this reaction is primarily due to the diethylsafranine monocarboxylic acid component since this compound gives a very good and specific reaction with these structures.

It was observed many years ago by T. B. Roberston that if one drop of a saturated solution of safranine was added to a solution of trypsin, a coloured precipitate was formed and subsequently it was demonstrated that this coloured precipitate had proteolytic activity. Then Marston demonstrated that other azo dyestuffs would react in a similar way, particularly neutral red which is a dimethyl-diaminotoluazine hydrochloride.

Marston suggested that these results indicated that the reaction of mitochondria with Janus green might signify that they contained proteolytic enzymes. Now, the concept of the staining reaction with Janus green has been pretty well proved by Lazarow and Cooperstein to indicate that mitochondria play an important part in cellular oxidations and that the production of a pink colour from the Janus green by the mitochondria is due to the DPN specific dehydrogenases.

The fact that Janus green B does not stain the mitochondria permanently green but that the green colour gradually becomes reduced to the pink and then to the colourless form has been known for many years. Joyet-Lavergne demonstrated more than twenty years ago that mitochondria contain an oxidase system which oxidizes cobaltous to cobaltic salts and that these latter stain the mitochondria green.

It was noted by Gatenby that in the snail, *Lymnaea,* the mitochondria are coloured yellow in the natural state presumably by a carotenoid pigment; also extracted lipomitochondria frequently have a yellow appearance which too is probably due to carotene. It is of interest in this connection that it has been demonstrated by the present author and by Joyet-Lavergne that mitochondria give a blue reaction with antimony trichloride in chloroform solution, an acknowledged reaction for vitamin A. Carotene is also a provitamin A, so these histochemical results indicate the presence of vitamin A in mitochondria.

Criticism of Joyet-Lavergne's results was made by Gomori because the former author had used alcohol as a fixative. The present author has, however, always applied antimony trichloride in chloroform solution direct to fresh unfixed tissues and this suggests that this result is a true one, in any case biochemical tests have now confirmed that mitochondria (prepared by homogenization and differential centrifugation of the cells) contain 27-32per cent of their weight of lipids and that 100 mg. of this lipid contains approximately 249-910 U.S.P. units of vitamin A.

JoyetLavergne has suggested that mitochondria contain a redox system in which vitamin A plays a part. A number of authors, Leblond, the present author and Giroud and his co-workers have demonstrated that mitochondria of some organs react with acetic acid-silver nitrate solution (which has been demonstrated as being a specific reagent for vitamin C) to give a positive reaction and this indicates that vitamin C may be present in them.

One has to accept the intracellular localization of vitamin C with a certain amount of discretion in view of the very destructive effect of this reagent on the cell cytoplasm. Electron-microscope studies in the present author's labouratory have demonstrated that this reagent has a most drastic effect on the ultrastructure of the cell and we cannot be sure that the localization of vitamin C in the mitochondria is, in fact, a real thing. It is of interest, however, that Chayen has found that the mitochondria of plant cells give this reaction very intensely and very specifically.

It may be that mitochondria in plant cells and some animal cells do, in fact, contain vitamin C, but we need further studies before this can be confirmed. Other vitamins appear to be present in mitochondria, members of the vitamin B complex have been found to be present, in some cases, in the form of coenzymes. The actual vitamins of the B complex recorded are vitamin B1 (thiamine), riboflavin, nicotinic acid (niacin), pantothenic acid and pyridoxine. However, although these vitamins are present in mitochondria, they are not present in any greater concentration than in the other parts of the cell so that they are not exclusively contained in these organelles.

These studies were made on cell homogenates and, of course, it is possible that the presence of the vitamins in the other fractions of the homogenate may be due to the fact that they have been leached out of the mitochondria by the saline solution which is used in the homogenization process in these particular experiments and further work would have to be done before we could be certain about this. The present author, Joyet-Lavergne and Giroud have demonstrated that mitochondria contain glutithione or protein-bound SH and this is further evidence that mitochondria can play an important part in oxidation-reduction mechanisms in the cell. It is of interest that mitochondria have been found in large quantities in the phloem cells of plants which are concerned with transport and may be concerned with this process. This suggestion is made in view of Conway's views that redox systems can play an important part in ion pumping. Another substance which was demonstrated histochemically in mitochondria by the present author, using the Schultz reaction, was cholesterol.

This was particularly evident in the mitochondria of cells of the adrenal cortex. Similar but less intense reaction was shown by the mitochondria of the liver and it has been demonstrated that isolated liver mitochondria contain about 2per cent cholesterol and it is possible they contain more in the adrenal cortex. Bensley has recorded the presence of a red pigment in the mitochondria and also in the submicroscopic particles of the liver cell. He believed that this pigment is derived from the oxidation of unsaturated fats and possibly the phospholipids of the mitochondria and this led him to suggest that in the liver cell the mitochondria in particular are, possibly, the seat of highly active oxidative processes which involve the metabolism of fats. The possible relation of mitochondria to oxidation-reduction processes was indicated by the publications of Ludford.

He demonstrated that, if methylene blue was added to tissue cultures, the mitochondria of the cells stained a brilliant blue colour but this could be inhibited by potassium cyanide. If the cells were exposed to a bright light, the blue colour was rapidly bleached. It is of interest that the mitochondria, although they have been demonstrated to contain a good deal of fat and lipid, do not give a positive reaction with Sudan III. They contain protein (quite a high proportion of it), but it is of interest that the earlier workers, using Millon's reagent, produced a negative reaction for protein. Subsequently Bensley and Gersh using a Millon's reagent of a different formula were able to obtain positive results from the mitochondria of many tissues and they found that they were particularly well stained in frozen, dried sections and particularly in those of *Amblystoma* liver.

The same authors demonstrated that, in undenatured, frozen, dried sections of *Amblystoma* liver, the mitochondria were destroyed if the sections were exposed to artificial gastric juice and artificial pancreatic juice. Despite all this chemical information most of which was in existence by the beginning of the 1940's, there was no certainty as to the function of mitochondria and it was not until 1946 and 1947 that the late George Hogeboom and his colleagues at Bethesda completely revolutionized our ideas of their function by demonstrating that the major proportion of the succinic dehydrogenase and an appreciable proportion of the cytochrome oxidase activity of the liver cell were present in the mitochondria. This at once suggested that these organelles were the major sites of aerobic respiration in the cell. It is of interest that as long ago as 1915, Dr. Kingsbury had suggested that mitochondria were concerned with the respiration of the cell. His reasons for this were largely due to his observations that anesthetics such as ether and chloroform, which depressed cellular respiration and the respiration of the animal in general, also broke up mitochondria in the cell. The work of Hogeboom and his colleagues was carried out on homogenates of liver which had been produced by grinding up liver with saline. This gave poor preservation of the form of the mitochondria and subsequently 0.8 *M* sucrose was used—this preserved the nature and form of the mitochondria very well.

In the beginning there was some doubt as to whether the material being assayed was in fact mitochondria or not, but when sucrose was used this doubt gradually disappeared. Eventually electron-microscope studies of the structure of the granules isolated by homogenization and differential centrifugation demonstrated beyond doubt that they contained the same structures as the mitochondria of normal cells.

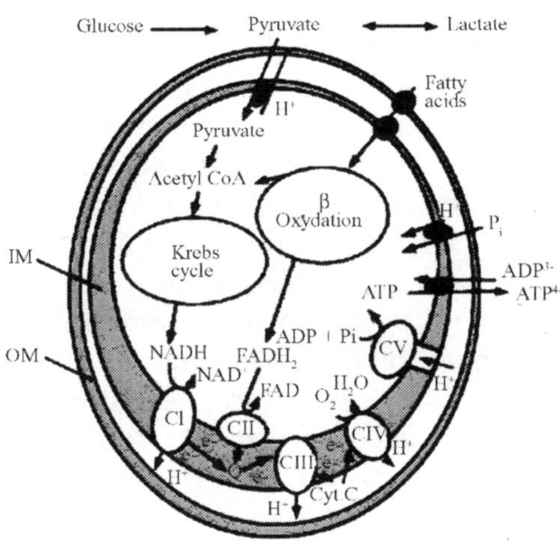

Fig. Respiratory Chain Enzymes

Another possible source of error, however, began to haunt the biochemists and was frequently verbalized by cytologists and this was that the mitochondria did not really contain the oxidative enzymes, mentioned in the foregoing, but that they were being absorbed or adsorbed by them from the homogenate. This possibility was negated at least in part by adding more enzyme to the homogenate and demonstrating that it could be recovered almost 100per cent from the supernatant and so was not taken up by the mitochondria. Thus the localization of the enzymes in the original cell was probably in the mitochondria. Subsequently, Kennedy and Lehninger demonstrated that in addition to containing cytochrome oxidase and succinic dehydrogenase, mitochondria also catalyze the condensation of pyruvic acid with oxaloacetic acid, the formation of succinic acid from α-ketoglutaric acid and the formation of malate from succinate.

These three reactions represent three steps of fundamental importance in the Krebs tricarboxylic acid cycle of which we will have more details shortly. Mitochondria also contain coenzyme I of which nicotinamide is an important constituent and cytochrome reductase which is a flavoprotein. These two enzymes are links between the Krebs tricarboxylic acid cycle and the cytochrome system and it, therefore, appears that mitochondria probably contain the whole enzymic equipment necessary for aerobic respiration of the cell.

In fact, it was originally demonstrated that a centrifugate of cells which included only nuclei and mitochondria were capable of carrying through the whole of the oxidation of glycogen to CO_2 and water; subsequently it was demonstrated that isolated mitochondria on their own could do this. Transaminase activity was subsequently demonstrated in mitochondria although in only the same concentration as in the rest of the cytoplasm. Transaminase is concerned with protein synthesis and it is of interest that pyruvic, oxaloacetic and α-ketoglutaric acids are the corresponding α-keto acids of the amino acids, alanine, aspartic and glutamic and need only to be transaminated to produce them.

Since the former compounds are themselves formed as intermediates in the course of the Krebs cycle, this indicates a mechanism whereby mitochondria might synthesize fresh protein material and so increase in size themselves or synthesize protein for other parts of the cell. This possibility is further extended by the fact that RNA is known to be present in varying amounts in mitochondria, although the exact amount is uncertain and there is a possibility of contamination of mitochondria with RNA from the rest of the homogenate when making such estimates. However, there is no doubt that mitochondria do have the equipment for synthesising protein although to what extent they do this *in vivo* is not known. All the fatty acid oxidase activity in the cell is also in the mitochondria and so is 80per cent of the octanoxidase activity.

Since these discoveries, an enormous amount of work has been done on the mitochondrial enzymes and this work was summarized by Hogeboom (unfortunately now deceased), Kuff and Schneider from the *National Institutes of Health* and has been published in Volume 6 of *The International Review of Cytology*, 1957. It may be of interest to record here other enzymes and compounds which they listed as being present in mitochondria. It is of interest too, that a considerable amount of ribonuclease and deoxyribonuclease are present in mitochondria—this seems unusual since only a very small amount of DNA and a relatively small amount of RNA are present in the mitochondria compared with other parts of the cell. However, it is possible that the bodies which contain RNAase and DNAase are not true mitochondria but are lysosomes; we will discuss these shortly. In the meantime, it is of interest to note that it was eventually demonstrated by fragmenting mitochondria and separating the membranes from the contents that cytochrome oxidase and succinic dehydrogenase activity were located primarily in the membranes. Subsequently it has been shown that the ability of the mitochondria to carry out the Krebs cycle and to combine this with oxidative phosphorylation is dependent on the presence of intact double membranes in the mitochondria. Now, it has been demonstrated and mentioned earlier that in the structure of mitochondria there is an outer membrane and an inner membrane to the mitochondrion, so that it is surrounded by a double membrane. Also, the inner membrane is folded inward to give elongated cristae and these cristae, because they consist of a fold of the inner membrane, are themselves double membranes.

There appears to be little difference biochemically between the double membrane structures of the cristae and the double membrane structure of the outer wall of the mitochondria. An important factor in the process of oxidative phosphorylation is the transport of electrons. Dr. David E. Green has described the process of electron transport as carried out fundamentally with the aid of an electron transport particle—this is a submitochondria particle. The latter can be isolated in a form so that it can no longer carry out the full citric acid cycle, but it can still carry out the process of oxidative phosphorylation which it performs by the oxidation of succinate and DPNH (reduced diphospho-pyridine nucleotide, a codehydrogenaste) with freshly prepared by homogenization and differential centrifugation and are then exposed to alternate freezing and thawing, their form undergoes considerable change.

First of all, the external membrane ruptures and some of the cristae break off and eventually the mitochondrion is completely disintegrated; however, fragments of double membrane structure are left which, according to D. E. Green, may be derived either from the cristae or from the outer envelope of the original mitochondrion. These double membrane fragments represent collections of Dr. Green's electron transport particles. If other methods are used for preparing these particles and the particles so produced do not show a double membrane structure, then oxidation of succinate and DPNH coupled with phos-phorylation does not occur.

Apparently the double membrane structure is essential for this process to take place. In the structures which do not have the double membrane structure, the capacity for electron transport is still present but oxidation phosphorylation cannot take place. About 35per cent of the dry weight of the electron transport particle of Green is composed of lipid and the lipid is concentrated in packets of lipoprotein. These are interspersed between two molecules of cytochrome or between a flavoprotein and cytochrome.

In other words, all the oxidation reduction members of the election transfer chain are connected to one another by lipoproteins. These lipoproteins, according to Green, act not only as structural devices, holding various enzymes together, but they also conatin a compound which can shuttle electrons back and forth so that then can pass from cytochrome through to flavoprotein and reverse. This compound, which Green and his colleagues have described, is a new coenzyme called coenzyme Q. It is a completely water insoluble benzoqtuinone derivative, which is connected in the lipoproteins of the electron transport particles and is related to vitamin K. Coenzyme Q can undergo oxidation and reduction in a similar fashion to cytochromes; for instance, the quinone part of the molecule can be reduced to become a hydroquinone and the hydroquinone can be reoxidized. It is pretty certain that this coenzyme plays an important part in the electron transfer chain of the particle.

The significance of the electron transport in these mitochondrial membranes may seem obscure but, as Green points out, one should think of this problem of electron transport in terms of its coupling—the fact that the process can be coupled to the esterification of inorganic phosphate. In this process a monophosphoric ester of the hydroquinone or, what Green described as a semiquinone form of coenzyme Q would be formed. This ester would then react with adenosine diphosphate by a process of transphosphorylation and ATP would be forme and simultaneously the semiquinone or hydroquinone would be oxidized to the quinone form in coenzyme Q by the ferric form of cytochrome c.

The production of ATP is, in fact, the reason for the existence of such a system. This explanation of the way in which the oxidative processes of the mitochondria actually are used to produce ATP by oxidative phosphorylation is an extremely fundamental discovery and Green and his co-workers are to be complimented very much on the extraordinary patience and care which they have put in over a large number of years to work out this very complex problem in mitochondrial physiology. In view of the studies by Green and his colleagues which have just been described, it is of interest to consider some of the studies of Martias.

This author has for some time put forward the view that there may exist in mitochondria two routes for the transport of hydrogen ions or electrons, respectively, between the pyridine nucleotides (DPN and TPN) and cytochrome c. He thought that only one of these would be linked with enzymes which brought about phosphorylation. The route which is best known leads from DPN to cytochrome c reductase and then to cytochrome c. Martias believes that this particular route is not one which is concerned with the formation of highenergy phosphates.

On the other hand, he proposed a second alternative route in which vitamin K (phylloquinone) is an important factor. Until recently this concept of an electron transport system involving vitamin K in mitochondria has not proved to be particularly popular, Martias and Strouff have finally isolated a phylloquinone reductase that is a flavoprotein having a number of similarities to cytochrome c reductase.

There is, however, a difference between them. Phylloquinone reductase and cytochrome c reductase can both react in the reduced state with phylloquinone, but with cytochrome c only cytochrome c reductase will have any effect. Vitamin K reductase has an extremely high activity even in relatively low concentrations. If we consider that there are two routes in mitochondria which can be used for electron transport, then the problem to be solved is what decides whether these electrons and hydrogen atoms are directed into either of these particular pathways.

Under normal circumstances they seem to be directed exclusively along the pathway which is coupled with oxidative phosphorylation and which leads via vitamin K reductase to vitamin K, cytochrome b and so through to cytochrome c. This pathway via vitamin K seems to be sensitive to any alteration of the internal structure of the mitochondria, possibly because some of its constituents, particularly the vitamin K, are associated with the double membranes of the cristae.

If their combination is disturbed, the electrons take the emergency route which is via cytochrome c reductase to cytochrome c. It is of interest in connection with this theory that dicoumarol is capable of decoupling oxidative phosphorylation from respiration and, since dicoumarol is known to be a vitamin K antagonist, its action is possibly explicable by the fact that it blocks the pathway between vitamin K and its reductase. It is also of interest that thyroxine is a hormone which causes uncoupling of respiration from oxidative phosphorylation.

Since it is known that, if thyroxine is added to mitochondria, it causes a swelling and change in the internal structure, this would be perhaps further evidence that the internal structure of the mitochondria is the one which is particularly concerned with the flow of electrons along the route leading to oxidative phosphorylation.

It may well be that hormones which affect respiration do so by affecting the structural nature of the mitochondria, apart from their effects on the endoplasmic reticulum which have already been mentioned. The precise relation between Green's coenzyme Q and Martias' vitamin K phylloquinone reductase is not absolutely clear, but it is very likely they are identical or closely related systems.

LYSOSOMES

De Duve in his studies on ultracentrifugation of cytoplasmic particles has demonstrated that some particles appear to contain acid phosphatase, cathepsin and ß-glucuronidase and ribonuclease, DNAase and cholinesterase. It is of interest that these compounds appear to be present in separate particles—they require higher centrifugal forces for sedimentation than do the cytochrome-oxidase bearing mitochondria. De Duve has pointed out that it is of special interest that such hydrolytic enzymes are in special particulate bodies differing from others of the cytoplasm. He is not clear as to how this should be interpreted but has pointed out that, if hydrolytic enzymes are free to act within the living cell, as, for instance, they can do in homogenates, they would interfere with the efficiency of the synthesis and may even affect the structural integrity of the cell; he then suggests that segregation of hydrolytic enzymes in this way is one method by means of which this activity is either kept in check or localized in specific parts of the cell.

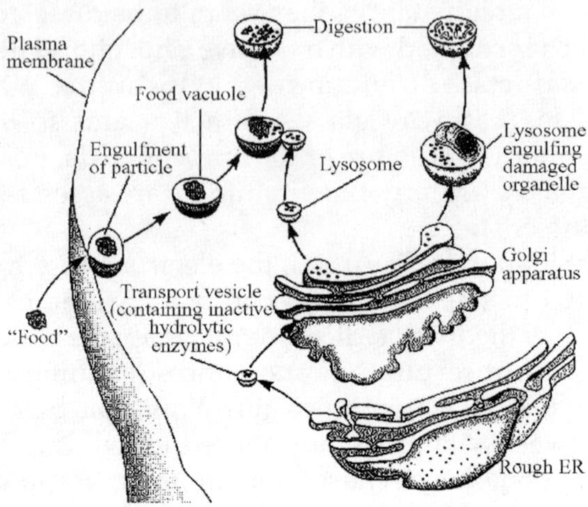

Fig. Lysosones.

De Duve once described these bodies as "suicide bags" and suggested that on the death of the cell these enzymes are released completely into the cytoplasm and play a part in the process of autolysis. The present author has suggested that these enzymes may be less effectively contained in the lysosomes in senescence and that this leakage may be partly responsible for the cellular process of aging. It is of interest that lysosomes have been demonstrated *in situ* with the electron microscope and stained to demonstrate their acid phosphatase activity. No typical mitochondrial structure can be seen in these bodies and they appear to be of a different nature from mitochondria.

MITOCHONDRIA: CELLULAR POWER PLANTS

Blink. Breathe. Wiggle your toes. These subtle movements—as well as the many chemical reactions that take place inside organelles—require vast amounts of cellular energy. The main energy source in your body is a small molecule called ATP, for adenosine triphosphate. ATP is made in organelles called mitochondria. Let's see if we can find some. They look like blimps about as long as pickup trucks but somewhat narrower. Oh, a few of them are over there. As we get nearer, you may hear a low whirring or humming sound, similar to that made by a power station. It's no coincidence. Just as power plants convert energy from fossil fuels or hydroelectric dams into electricity, mitochondria convert energy from your food into ATP.

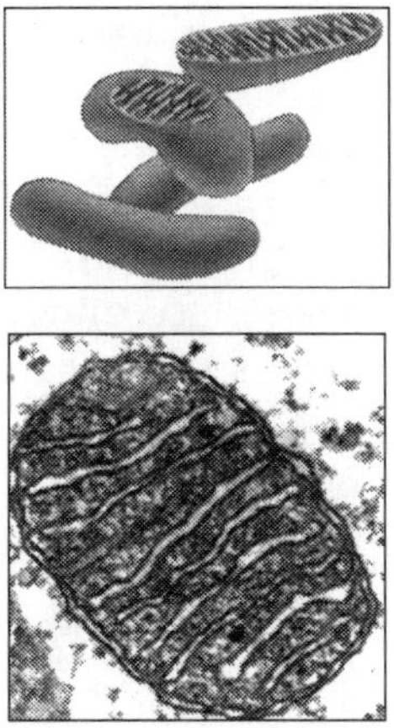

Fig. Mitochondrion

Like all other organelles, mitochondria are encased in an outer membrane. But they also have an inner membrane. Remarkably, this inner membrane is four or five times larger than the outer membrane. So, to fit inside the organelle, it doubles over in many places, extending long, fingerlike folds into the center of the organelle. These folds serve an important function: They dramatically increase the surface area available to the cell machinery that makes ATP. In other words, they vastly increase the ATP–production capacity of mitochondria. The mazelike space inside mitochondria is filled with a strong brew of hundreds of enzymes, DNA (mitochondria are the only organelles to have their own genetic material), special mitochondrial ribosomes, and other molecules necessary to turn on mitochondrial genes.

	Actual Size (Average)	**Perceived Size when Magnified 3 Million Times**
Cell diameter	30 micrometers	300 feet
Nucleus diameter	5 micrometers	50 feet
Mitochondrion length	Typically 1–2 micrometers, but can be up to 7 micrometers long	18 feet
Lysosome diameter	50–3,000 nanometers	5 inches to 30 feet
Ribosome diameter	20–30 nanometers	2–3 inches
Microtubule width	25 nanometers	3 inches
Intermediate filament width	10 nanometers	1.2 inches
Actin filament width	5–9 nanometers	0.5–1 inch

CELL DIVISION

From the extensive investigations on the mechanism of cell division it must be concluded that there is as yet no full understanding of the physical and chemical mechanisms which bring forth this process. There is agreement however that we are confronted with a series of events which together culminate in the phenomena of membrane formation and cell division.

The initial process, formation of fertilization membrane and entrance of the spermatozoon into the ovum with no nuclear changes, may be called the *anaerobic phase*, because it may take place under strictly anaerobic conditions. If energy is required for this process it must be obtained from glycolysis.

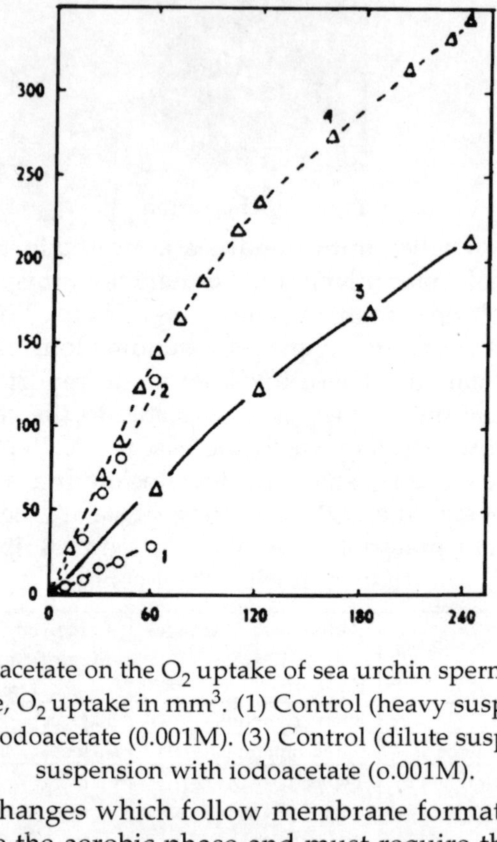

Fig. Effect of iodoacetate on the O_2 uptake of sea urchin sperm. Abscissa, time in minutes. Ordinate, O_2 uptake in mm^3. (1) Control (heavy suspension). (2) Heavy suspension with iodoacetate (0.001M). (3) Control (dilute suspension). (4) Dilute suspension with iodoacetate (o.001M).

The nuclear changes which follow membrane formation and end in cell division constitute the aerobic phase and must require the energy provided by oxidative reactions. The increased utilization of glycogen accompanied by the increased utilization of pyruvate may provide the energy necessary for the process of cell division.

Furthermore, the introduction during the cell division process of three of four oxidative steps (the series of oxidations carried on by the cytochrome system) in the series of reactions from substrate to molecular oxygen will increase the efficiency of the reactions so that for the same amount of O_2 utilized there will be a more efficient utilization of energy. Whether carbohydrate is metabolized entirely via glycolysis, or is oxidized through the pathway of hexosemonophosphate oxidation as suggested by Örström and Lindberg is not known. Neither is the nature of the acid formed on cell division known. Some of it may be acetic acid formed on oxidation of pyruvic acid. Whether acetate is utilized or not is not known. If it is, the pathway of its oxidation must be different from that of most animal tissues because neither succinate nor α-keto-glutarate is oxidized by sea urchin eggs. Fluoroacetate had no effect on the O_2 uptake.

Table. Utilization of Pyruvate by Arbacia Eggs—Influence of Fertilization

Date (2001)	*Fertilization (%)*	*Duration (hrs.)*	*Pyruvate utilized per gm. dry weight per hr. (micrograms)*
August 17	80	1	400
August 19	75	2	500
August 21	80	5	345
August 23	88	3	480
August 24	80	0.5	500
August 21	none	1	60
August 23	none	5	70
August 24	none	2	80
August 17	none	1	50
August 19	none	3	69

CELL GROWTH

Our knowledge of the enzymatic reactions which take part in cellular growth is meager. In general cellular growth, new protein formation, takes place at the expense of aerobic oxidations except in those cells. deprived of cytochromes. Voegtlin, Hammett, and Rapkine have for a long time championed the importance of sulfhydryl groups in cell growth, and evidence in support of this opinion is good. They have been found increased in concentration in growing cells and growth was inhibited on addition of substances which combine with —SH groups. It has already been pointed out that there are a large number of enzymes which require the presence of —SH groups in their protein moiety for activation.

Destruction of these groups by oxidation (x-rays, O_2, Cu), alkylations (CH_2 ICOOH, CH_2 ICOONH$_2$, mustard, nitrogen mustard), or mercaptide formation (Hg, Pb, Cd, Zn, Bi, As, V, Sb) will produce enzyme inhibitions, an inhibition which affects such high-energy yielding reactions as pyruvate oxidase, α-ketoglutarate oxidase, succinoxidase, triosephosphate oxidase, adenosinetriphosphatase, malate oxidase. Sluggish oxidation-reduction systems, mostly in their reduced state (glutathione, ascorbic acid), have also been found abundant in cells during growth and development. These systems must have for function that of keeping the —SH groups of the enzymes in the reduced, physiologically active state.

It must be recalled that ionizing radiations, halogenated alkylamines, and mustard gas inhibit mitosis. Furthermore Hevesy has found that the synthesis of nucleoproteins (which take part in mitosis) is inhibited by x-rays. Under certain conditions enzymes like adenosinetriphosphatase from aged myosin are so susceptible to ionizing radiations that amounts as small as 1r produce inhibition.

The inhibiting effect of x-rays may well indicate that some enzyme necessary for the synthesis of nucleoproteins requires —SH groups for its activity, groups which are oxidized by the OH and O_2 H radicals, and H_2O_2 produced during the irradiation of water. The inhibiting effect of mustard gas and of nitrogen mustard might also be due to combination of these substances with the —SH groups. The fact that lewisite produces no effect is no serious objection, because Barron*et al.* have shown that d-amino acid oxidase, a sulfhydryl enzyme, was not inhibited by lewisite.

The close interrelation between the metabolism of carbohydrates, fats, and proteins has shown that for the synthesis of protoplasm energy is obtained by the aerobic oxidation of carbohydrate. The preponderant part of synthesis reactions over breakdown reactions in growing cells must be due not only to the increased formation of enzymes but also to the distribution of the regulating mechanisms which not only modify the rate of reactions but are also able to change the direction of these reactions.

Table. The Effect of Very Low Doses of X-Ray on Myosin

	Phosphorus liberated in 10 min. at 38° (micrograms)		
Dosage (r)	*Treated*	*Control*	*Inhibition (%)*
10	5.7	11.9	52
5	5.4	11.9	55
1	8.4	11.9	30

The study of these regulating mechanisms which channel the multiple enzymatic reactions towards different pathways will clarify the role of enzymes in growth processes and will help us in understanding the change from normal controlled growth to the uncontrolled growth of cancer.

DEFFERENTIAL GROWTH BY BAUL WEISS

THE appearance of different properties in initially equivalent cells is called "differentiation." Growth being only one cellular attribute among many, "differential growth" is to be subsumed under the larger topic of "differentiation," and our discussion therefore will have to centre on the latter. Since the accent of this book lies more on prospects than on retrospects, It is intend to concentrate on the gaps in our concepts and knowledge of differentiation and point to possible ways of filling them rather than dwell on past achievements, which, though impressive in themselves, are dwarfed by the task that remains to be accomplished.

Biology has discovered the strength it can derive from a more rigourous analysis of biological phenomena by the tools and standards of the physical sciences, and if the study of development is to share the new light with its more alert sister branches, such as physiology, genetics and immunology, to name only a few, it will have to submit to the same reorientation that has proved so beneficial with the others. This reorientation is in the direction of greater precision, objectivity, and consistency in the description and interpretation of phenomena, preferably in terms of physical and chemical order. It implies the adoption not only of the tools of the physical sciences, but above all of their disciplined methodology. Analytical embryology is still full of problems which it can only settle within its own right and by its own techniques. But in doing this it must apply standards, criteria, and methods of procedure as rigourous as those of the physical sciences. It must strive to replace the abstract formalism and verbalism of its groping infancy by a mature realism, in which a term stands as symbol for a known or knowable entity or relation, rather than as a soothing device of convenient elasticity to pretend knowledge where there is none.

We must not allow its analytical principles to be contaminated by relapses into animistic mythology, as we are doing when we charge the unresolved residue of meticulous determinations of physico-chemical properties to "organizers," "inductors," and similar ill-descript agents. Labels must no longer be allowed to pass for content, nor generalities for explanations. What we need most in entering the era ahead is the will to drive the analysis of developmental phenomena far beyond the mark where we used to stop to coin a term, and the determination to become as exact, specific, objective, and realistic in the formulation of our problems and in the description of our observations and results as has become the rule in the physical sciences. This is what we mean by reorientation. It is just as important as is the gathering of more data. With this in mind, we turn to the question which forms the core of my assignment: What is differentiation?

WHAT IS DEFFERENTIATION?

By the standards just advocated, this question can immediately be recognized as far too indefinite to be answered by anything but another generality. For "differentiation" refers not to a single circumscribed natural phenomenon, but is a summary name for a heterogeneous collection of notions about a variety of events associated with development.Consequently, the various statements and generalizations which have been made in the past concerning "differentiation" and "dedifferentiation" are equally heterogeneous and full of contradictions. Only a few authors have deemed it necessary to be explicit about terms of such general currency. A definition of "differentiation" is partly a matter of objective description, partly of plain logic. We shall deal with the logical portion first, because it has been more persistently ignored.

CHARACTERS VS. PROCESSES

One source of confusion is the tradition of confounding the process of differentiation with its criteria. We usually tell differentiation by its more conspicuous products; *e.g.* melanin granules in a pigment cell, myofibrils in a muscle fibre, secretion granules in a gland cell. Our criteria are predominantly morphological and we judge mostly by appearances. In the examples just mentioned, our judgement happens to be well founded.

It becomes dubious when based merely on superficial differences of shape, arrangement, or size, and it misleads us completely when we try to reverse the argument and take the absence of obvious differences for evidence of lack of differentiation. To appraise a cell correctly we must take into account all of its properties, not just the ones that impress us optically, and to appreciate the development of these properties we must bear in mind that we are dealing with material systems in space and time, in which whatever is present at any one stage is the result of antecedent *processes*. Thus the conventional "characters" by which we distinguish differentiation are in reality only incidental manufacturing samples of the cellular production plant.

And just as one kind of industrial plant differs from another not only by its different end products but by the whole machinery and processing method turning out those products, so a differentiated cell of one kind differs from that of another kind not only in its final equipment (pigment, fibrils, secretions, etc.) but in the basic mechanisms through which that equipment has been produced. Tracing pigment cells, muscle cells, or blood cells back into earlier phases of their development, we find, or postulate, that they contain then, instead of the terminal products of melanin, myosin, or hemoglobin, half-finished "precursors." But if we go back still further, beyond the "precursors of precursors," the future products lose their identity and we are faced with nothing but the plant capable of turning them out, and perhaps some of the ingredients.

Now, evidently a plant designed to turn out chiefly melanin must be different from one designed to make contractile myofibrils. We express this fact in the names "melanoblast" and "myoblast" given to those early stages, and we generally concede that they are intrinsically different, although the differences are no longer demonstrable by ordinary microscopic or chemical techniques. By further refinements of technique we can extend the range of our discriminative powers; structurally, by the ultramicroscope, polarization microscope, electron microscope and x-ray diffraction; chemically, by microtests (including cytological stains), spectroscopy, microincineration, etc. These techniques will undoubtedly reveal criteria of differentiation much earlier in the embryonic history than was possible with the cruder classical tools of observation.

This is a line of investigation worthy of vigourous prosecution. But we must expect that it too will come to an end short of retracing the whole ontogenetic course of a given cell strain, for we are still essentially concerned with tracing "products" rather than the production processes that lie behind. Once we can detect the "product," most of the story of its production is over and we have certainly missed its essential beginnings. It looks as if the analytical methods at hand are of no avail.

Preformistic theory would deny the validity of our dilemma. It would assert that all distinctions of later stages have existed, though less manifestly, from the beginnings of ontogeny. But experimental embryology has thoroughly discredited this concept by proving that mature cells which differ profoundly in their physico-chemical constitution may be derived from cells that are demonstrably equivalent and, to all practical purposes, identical. When and by what means have their developmental courses then become divergent? This is the crucial question of cytodifferentiation. Fortunately, though analytical techniques fail us, we can answer it by a test that might be called "behavioural."

This is where we must draw on logics rather than gadgets. Suppose we are to compare two closed material systems, whether they be molecules or cells or organisms, about whose character and composition we know nothing. How are we to tell whether they are the same or different? The only way of telling is by watching their behaviour. If both behave identically under identical conditions, we decide that they are alike. However, if both, again *under identical conditions, behave differently*, then we must conclude that *they have been intrinsically different* from the very beginning of the test. Against the cogency of this postulate even the most persistent failures to detect more tangible signs of difference would prevail nothing. Two seeds giving rise to different plants in the same soil, or two eggs giving rise to different animals in the same pond, would have to be acknowledged as intrinsically different even if they were otherwise indistinguishable.

Similarly, if two cells under identical conditions behave and respond differently in any respect (*e.g.* movement, growth rate, chemical activity, sensitivity to radiation or drugs, etc.), then they are different, which implies that if they have descended from a common source of identical cells, they must have become different, *i.e.* have gone through a process of differentiation, no matter whether they show it by morphological signs or not. The behavioural test is as compelling as any morphological test, and far more pertinent; for morphological criteria are but residues and convenient indicators of prior activities. Early differences of this type, which are not immediately recognizable and can only be deduced from later formative activities, are commonly described as differences of "potency." If this were to imply that they are virtual rather than real (in the sense of Driesch), then it would mean reinstating the old "vital spirits" under another name.

Actually there can be no doubt that we are dealing with differences of equipment and conditions which are real, yet too subtle to be detectable by present techniques. In conclusion, we propose to define "differentiation" as the unidirectional changes in the character of cells and cell generations during their life history, transformations by which they become increasingly different from their own former selves, their parent cells, and the cells of other strains which have taken divergent courses. And, as we have already intimated, "character" is to include the totality of reactions of which the cell is capable at a given stage. In any given situation, only a fraction of these possible reactions will be realized, and only a small part of this fraction will become externally recognizable.

MODULATION

We have said that cells that look alike may be intrinsically very different. The opposite is equally true, namely, cells may look very different and yet be essentially alike in character. Let me cite a striking example from my own experience, the vagaries of the adult sheath cell of Schwann of adult mammalian nerve. In the embryonic stage these cells are long spindles. They then envelop the nerve axons as thin protoplasmic cylinders. When axons degenerate, as after nerve section, the sheath cells regain mass, become spindly again, and glide out from the nerve tubes.

After nerve regeneration they resume their original sheathing position and corresponding cylindrical shape. But they can vary even much more than just shuttling back and forth between these two states. Embryonic Schwann cells, reared in tissue culture along the interface of cover glass and fluid medium, tend to transform from the spindly type into a flat round form of the appearance and behaviour of macrophages. A repetition of the experiments with adult nerve proved that the fully mature Schwann cell is capable of the same transformations and more. Plate illustrates a series of cell forms obtained *in vitro* from Schwann cells of peripheral nerve.

They include tandem strands of filamentous sheath cells, ramified astrocyte-like forms, giant macrophages with ruffled membrane in active phagocytosis, monocyte-like round cells with clear ectoplasm and eccentric nucleus, and foamy round cells whose vitality is proven by their mitotic activity. Each one of these phases occurs under certain definable conditions, among which the fine-structural constitution of the medium seems to play a predominating role. Some of the transformations are easily reversible, others have as yet been observed in one direction only. Since they can be followed directly as they take place in the cultures, valuable insight into the processes underlying the morphological results has been obtained. There was evidence of significant changes in the distribution of the cell content, size of the ectoplasmic border, constitution, elasticity, rigidity, mobility, etc., of the cell surface, average number and shape of pseudopodia, mode and rate of locomotion, viscosity, vacuolization, adhesiveness, orientation of fibrillar plasma constituents (as indicated by silver impregnation), fat production, distribution of nuclear materials, shape and hydration of nucleus and nucleoli, phagocytic power, and other associated properties.

In short, these cells, in response to different environments, have undergone rather profound reorganizations of the kind unhesitatingly classified as "differentiation" or "metaplasia" by a nomenclature based on sheer appearances. But is this type of change directly comparable to the progressive cytodifferentiation during ontogeny of which we have spoken before? It evidently is not, and it is in this place that facts supplement logic. As we mentioned before, many of the observed changes are reversible if the cell is returned to its prior environment. There are many more cell types of the mature organism which behave similarly in that their behaviour and shape change reversibly with their environment.

Changes in the nutrient and hormone composition of the internal milieu, sublethal toxic agents, inflammatory processes, seasonal fluctuations, denervation, and in fact, on a more moderate scale, practically all ordinary physiological stimulations, cause a more or less thorough reshuffling of the cell content with consequent change in cell behaviour, and often in cell form. The important point to keep in mind is that in all these instances the cell can revert to its original state when the original environmental conditions are restored. Evidently the cell passes through these transmutations with no alteration, either gain or loss, of its basic equipment. At the end of a cyclic change it turns out to be the same cell that it was before, and it is quite beside the point that sometimes the reversal may not occur in the same cell, but in one of its descendants which then prove to be "chips off the old block."

There is no reason why the reversible changes observed in our Schwann cells should be regarded in any different light than, let us say, the contractions and expansions of a melanophore in response to cyclic nervous or humoral stimuli, or the change in the secretory state (and associated morphology) of a hormone-sensitive cell in response to varying concentrations of the respective hormone. They all are merely expressions of how a given cell can react to a variety of conditions—its "reaction repertory," based on its material constitution such as it is at the time. The living insight into the processes underlying the morphological results has been obtained. There was evidence of significant changes in the distribution of the cell content, size of the ectoplasmic border, constitution, elasticity, rigidity, mobility, etc., of the cell surface, average number and shape of pseudopodia, mode and rate of locomotion, viscosity, vacuolization, adhesiveness, orientation of fibrillar plasma constituents (as indicated by silver impregnation), fat production, distribution of nuclear materials, shape and hydration of nucleus and nucleoli, phagocytic power, and other associated properties. In short, these cells, in response to different environments, have undergone rather profound reorganizations of the kind unhesitatingly classified as "differentiation" or "metaplasia" by a nomenclature based on sheer appearances.

But is this type of change directly comparable to the progressive cytodifferentiation during ontogeny of which we have spoken before? It evidently is not, and it is in this place that facts supplement logic. As we mentioned before, many of the observed changes are reversible if the cell is returned to its prior environment. There are many more cell types of the mature organism which behave similarly in that their behaviour and shape change reversibly with their environment.

Changes in the nutrient and hormone composition of the internal milieu, sublethal toxic agents, inflammatory processes, seasonal fluctuations, denervation, and in fact, on a more moderate scale, practically all ordinary physiological stimulations, cause a more or less thorough reshuffling of the cell content with consequent change in cell behaviour, and often in cell form. The important point to keep in mind is that in all these instances the cell can revert to its original state when the original environmental conditions are restored.Evidently the cell passes through these transmutations with no alteration, either gain or loss, of its basic equipment.

At the end of a cyclic change it turns out to be the same cell that it was before, and it is quite beside the point that sometimes the reversal may not occur in the same cell, but in one of its descendants which then prove to be "chips off the old block." There is no reason why the reversible changes observed in our Schwann cells should be regarded in any different light than, let us say, the contractions and expansions of a melanophore in response to cyclic nervous or humoral stimuli, or the change in the secretory state (and associated morphology) of a hormone-sensitive cell in response to varying concentrations of the respective hormone.

They all are merely expressions of how a given cell can react to a variety of conditions—its "reaction repertory," based on its material constitution such as it is at the time. The living ments, in each other's places. This can be done by the techniques of explantation (tissue culture) and of transplantation, respectively. The answer has been unequivocal. Cells taken from different tissues of late embryonic or mature organisms and transferred into standard extraneous media offering nutrition, protection, and support lose many of their specialized aspects, *i.e.* criteria by which we used to tell their "differentiation." Spectacular as their differences may have been during residence in the body, they soon come to look and behave much more alike, in adaptation to the common environment. In the early days of tissue culture this simplification of appearance was actually often interpreted as a sign of "dedifferentiation," in the sense of a return not only to a more primitive but to a veritably primordial state.

This was again judgement by appearances with all its pitfalls. Yet later work showed clearly that the explanted cells, in spite of their similar disguise, retain distinguishing features of fundamental nature which they pass on unaltered to their descendant cells for indefinite numbers of generations. As a glance at Plate will illustrate, Schwann cells and endoneurial connective tissue cells, while assuming both the aspect of common "mesenchym," *i.e.* spindle shape and amoeboid activity, can be well distinguished by their sizes, nuclear characteristics, staining properties, affiliations, and other criteria revealed only by subtler analysis.

In other words, the parallel modifications which the cell strains undergo on transfer to tissue culture are merely modulations carried out by variously "differentiated" cells that otherwise preserve their basic distinctions. Their distinctiveness persists even during such further modulations in culture as the transformation to the macrophage stage. This transformation has thus far been observed in cultures of various connective tissues, Schwann cells, and muscle. They all can give rise under similar conditions to a highly phagocytic amoeboid cell with hydrated nucleus, dense entoplasm, and a more hyaline ruffled border, *i.e.* macrophage-like characters.

But in doing so they also retain the signs of their descendancy, so that a "macrophage" resulting from the conversion of an endoneurial cell can usually be clearly distinguished from one that has come from a Schwann cell. Tissue culture thus does not really abolish character differences established during ontogeny. This conclusion is fully borne out by the critical litreature on the subject. It is strikingly confirmed by experiments in which cells that had lost their more specialized aspects during intensive proliferation in tissue culture were again given opportunity to produce their specialties, either by suppressing further proliferation *in vitro* or by transplanting them back into an organism.

Under such conditions they were essentially true to the kind from which they stemmed: those derived from pigment epithelium would resume the production of pigment granules; those from thyroid, of colloid; those from gut, of digestive enzymes. Cultured cells returned to an organism, but into atypical locations, moreover, showed no tendency to differentiate in conformity with their new environment, proving that the culture period had not restored to them the ability of younger cells to switch into a new local course of differentiation. All this adds up to the realization that the measure of true differentiation is the absence of true dedifferentiation, if dedifferentiation connotes not the mere assumption of a semblance of primitivity but the actual return to a state of wider potency, *i.e.* the recuperation of something that had ostensibly been lost. By the testimony of the reported and similar experiments, vertebrate cells are incapable of such true dedifferentiation; hence, for them, true differentiation is the rule.

PLANT CELL MODEL

Biotechnology is name given to the methods and techniques that involve the use of living organisms like bacteria, yeast, plant cells etc or their parts or products as tools (for example, genes and enzymes). They are used in a number of fields: food processing, agriculture, pharmaceutics, and medicine, among others. Plant tissue culture can be defined as culture of plant seeds, organs, explants, tissues, cells, or protoplasts on nutrient media under sterile conditions. The science of plant tissue culture takes its roots from path breaking research in botany like discovery of cell followed by propounding of cell theory. In 1839, Schleiden and Schwann proposed that cell is the basic unit of organisms. They visualized that cell is capable of autonomy and therefore it should be possible for each cell if given an environment to regenerate into whole plant. Based on this premise, in 1902, a German physiologist, Gottlieb Haberlandt developed the concept of *in vitro* cell culture.

He isolated single fully differentiated individual plant cells from different plant species like palisade cells from leaves of Laminum purpureum, glandular hair of Pulmonaria and pith cells from petioles of Eicchornia crassiples etc and was first to culture them in Knop's salt solution enriched with glucose. In his cultures, cells increased in size, accumulated starch but failed to divide. Therefore, Haberlandt's prediction failed that the cultured plant cells could grow, divide and develop into embryo and then to whole plant. This potential of a cell is known as totipotency, a term coined by Steward in 1968. Despite lack of success, Haberlandt made several predictions about the requirements in media in experimental conditions which could possibly induce cell division, proliferation and embryo induction. G Haberlandt is thus regarded as father of tissue culture.

Taking cue from Haberlandt's failure, Hannig chose embryogenic tissue to culture. He excised nearly mature embryos from seeds of several species of crucifers and successfully grew them to maturity on mineral salts and sugar solution. In 1908, Simon regenerated callus, buds and roots from Poplar stem segments and established the basis for callus culture. For about next 30 years, there was very little further progress in cell culture research. Within this period, an innovative approach to tissue culture using meristematic cells like root and stem tips was reported by Kolte and Robbins working independently. All these research attempts involving culture of isolated cells, root tips or stem tips ended in development of calluses. There were two objectives to be achieved before putting Haberlandt's prediction to fruition. First, to make the callus obtained from the explants to proliferate endlessly and second to induce these regenerated calluses to undergo organogenesis and form whole plants. It was in 1930s, when progress in plant tissue culture accelerated rapidly owing to an important discovery that vitamin B and natural auxins were necessary for the growth of isolated tissues containing meristems. This breakthrough came from White who reported that not only could cultured tomato root tips grow but could be repeatedly subcultured to fresh medium of inorganic salts supplemented with yeast extract. He later replaced YE by vitamin B namely pyridoxine, thiamine and proved their growth promoting effect.

DESCRIPTION

Plant cells are the cells found in plants comprising sub cellular organelles. Following is a plant cell model for kids that will give your child a visual of the plant cell and a better understanding of its various parts.

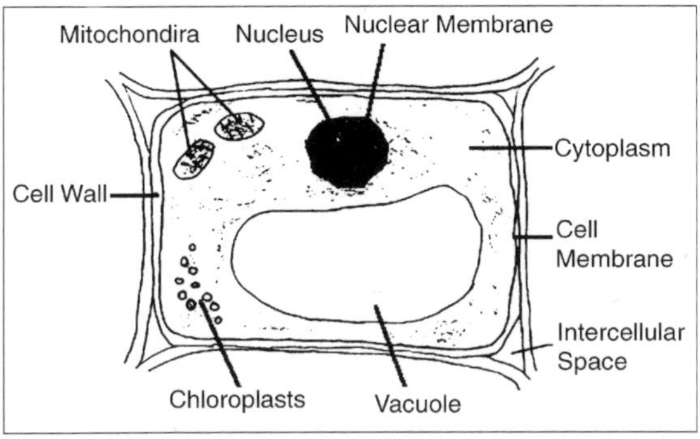

The cells are the basic units of life that work together to perform life sustaining functions in both animal and plant worlds. Plant cells are eukaryotic cells having thick and rigid cell walls. Eukaryotic cells are cells that contain complex structures enclosed within membranes called nucleus or nucleus envelop, within which the genetic material is present. These cells are present in almost all living organisms including animals, plants and fungi.

PLANT CELL MODEL AND PARTS

Plant cells differ from animal cells in that they have three different structures known as cell wall, vacuoles and plastids. However, they lack centrioles and intermediate filaments, which are present in animal cells. Read more on plant cell structure and parts.

Cell Membrane

It permits the waste material to exit the cell and regulates the movement of materials in and out of the cell. It also acts as a barrier between the inside of the cell and the outside, so that the chemical conditions on both sides can be different.

Cell Wall

It provides rigid structural support to the cell from which structures, like, leaves and stems are produced. It allows the circulation and distribution of water, minerals and nutrients in and outside the cell. The wall also controls the presence of pathogen microbes and development of tissues in the cell.

Golgi Body

It stores, packages, and circulates the lipids and proteins made in the endoplasmic reticulum. It stores the proteins into packages called vesicles.

Rough Endoplasmic Reticulum

It synthesizes and exports proteins and glycoproteins throughout the cell.

Lysosomes

These organelles digest food by breaking down larger molecules into smaller ones, using special proteins.

Cytoplasm

It helps in maintaining the cell shape. Along with this, it plays an important role in the internal movement of cell organelles, cell mobility and muscle fibre contraction. It also distributes oxygen and nutrients to different parts of the cell.

Nucleolus

It is the most prominent structure in the nucleus wherein ribosomes are made.

Vacuole

The plant cell contains a large, single vacuole (an enclosed compartment), which is used to store water, compounds and minerals that help in plant growth.

Ribosomes

These are packets of RNA (Ribonucleic Acid) and are called as protein builders or synthesizers of the cell.

Chloroplasts

They are the food producers of the cell, containing chlorophyll, the green pigment essential for photosynthesis. Their main function is to generate sugars and starches.

Nucleus

It's the most important part of the cell, comprising chromosomes, *i.e.* structures made up of genetic information that helps in cell growth and reproduction. Read more on cell nucleus: structure and functions.

Nuclear Envelope

It's an enclosure that surrounds nucleus and its contents. Unlike cell membrane, which has pores and spaces for RNA and proteins to pass through, it keeps the chromatin and nucleolus inside the nucleus.

Smooth Endoplasmic Reticulum

It's main function is to package proteins for transport, synthesize membrane phosolipids, and secrete calcium. It also performs transformation of bile pigments, glycogenolysis (the breakdown of glycogen), and detoxification of different drugs and chemical agents.

Mitochondria

It provides energy to the cell by combining sugar molecules with oxygen to generate carbon dioxide and water.

Amylosplast

It's an organelle present in some plant cells that stores starch.

Druse Crystal

It's a granular type of crystal found in plant vacuoles. It is composed of calcium oxalate and is considered to deter herbivory.

Centrosome

Also known as Microtubule Organizing Center, it's a region in the cell where microtubules are produced that perform a variety of functions ranging from transportation to structural support. Though both plant and animal cell centrosomes play similar roles during cell division, plant cell centrosomes are simpler and do not contain centrioles.

Peroxisomes

These are membrane bound packets of oxidative enzymes that convert fatty acids into sugar and assist chloroplasts in photo-respiration.

Golgi Vesicles

These vesicles transfer proteins and lipids to different parts of the cells.

PLANT CELL MODEL

Following are some simple instructions or plant cell model ideas, to construct a model of a plant cell:

- Enumerate and collect all possible materials required for the model and look out for the different cell parts in a biology textbook.
- Use a heavy sturdy wooden base to construct the base of the model so that it can hold the weight and facilitate presentation of the model.
- Construct the cell nucleus and ensure that it's round in shape. Paint the

nucleus in one colour so that it can be identified as a distinct cell part.

- With the help of a dowel rod or thick, straight stick, connect the nucleus to the base of the model. For this, drill a hole into the wooden base, stick the dowel rod into the hole and connect the nucleus to the rod.
- The cell wall and cell membrane of the plant cell model should be rectangular in shape. The outer wall is the cell wall, which should be made of a strong material like hard plastic or wood, and the cell membrane can be made of cellophane or thin plastic. Now attach both the walls to the nucleus and the model base.
- Add various other key components of the plant cell to the model. Like, for chlorophyll, use a piece of green fabric stuffed with cotton. Sew it and attach it to the model.
- Finally, prepare a report on the various plant cell parts and their corresponding functions.

5

DNA Replication within the Cell

For a cell to divide, it must first replicate its DNA. This process is initiated at particular points within the DNA, known as "origins", which are targeted by proteins that separate the two strands and initiate DNA synthesis. Origins contain DNA sequences recognized by replication initiator proteins (eg. dnaA in *E coli'* and the Origin Recognition Complex in yeast).

These initiator proteins recruit other proteins to separate the two strands and initiate replication forks. Initiator proteins recruit other proteins to separate the DNA strands at the origin, forming a bubble. Origins tend to be "AT-rich" (rich in adenine and thymine bases) to assist this process because A-T base pairs have two hydrogen bonds (rather than the three formed in a C-G pair)—strands rich in these nucleotides are generally easier to separate. Once strands are separated, RNA primers are created on the template strands and DNA polymerase extends these to create newly synthesized DNA. As DNA synthesis continues, the original DNA strands continue to unwind on each side of the bubble, forming replication forks. In bacteria, which have a single origin of replication on their circular chromosome, this process eventually creates a "theta structure". In contrast, eukaryotes have longer linear chromosomes and initiate replication at multiple origins within these.

THE REPLICATION FORK

The replication fork is a structure which forms when DNA is being replicated. It is created through the action of helicase, which breaks the hydrogen bonds holding the two DNA strands together. The resulting structure has two branching "prongs", each one made up of a single strand of DNA.

Leading Strand Synthesis

In DNA replication, the leading strand is defined as the new DNA strand at the replication fork that is synthesized in the 5′→ 3' direction in a continuous manner. When the enzyme helicase unwinds DNA, two single stranded regions of DNA (the "replication fork") form. On the leading strand DNA polymerase III is able to synthesize DNA using the free 3' OH group donated by a single RNA primer and continuous synthesis occurs in the direction in which the replication fork is moving.

Lagging Strand Synthesis

The lagging strand is the DNA strand at the opposite side of the replication fork from the leading strand, running in the 3' to 5' direction. Because DNA polymerase cannot synthesize in the 3'→5' direction, the lagging strand is synthesized in short segments known as Okazaki fragments. Along the lagging strand's template, primase builds RNA primers in short bursts. DNA polymerases are then able to use the free 3' OH groups on the RNA primers to synthesize DNA in direction. The RNA fragments are then removed (different mechanisms are used in eukaryotes and prokaryotes) and new deoxyribonucleotides are added to fill the gaps where the RNA was present. DNA ligase then joins the deoxyribonucleotides together, completing the synthesis of the lagging strand.

DYNAMICS AT THE REPLICATION FORK

As helicase unwinds DNA at the replication fork, the DNA ahead is forced to rotate. This process results in a build-up of twists in the DNA ahead. This build-up would form a resistance that would eventually halt the progress of the replication fork. DNA topoisomerases are enzymes that solve these physical problems in the coiling of DNA. Topoisomerase I cuts a single backbone on the DNA, enabling the strands to swivel around each other to remove the build-up of twists.

Topoisomerase II cuts both backbones, enabling one double-stranded DNA to pass through another, thereby removing knots and entanglements that can form within and between DNA molecules. Bare single-stranded DNA has a tendency to fold back upon itself and form secondary structures; these structures can interfere with the movement of DNA polymerase. To prevent this, single-strand binding proteins bind to the DNA until a second strand is synthesized, preventing secondary structure formation.

Clamp proteins form a sliding clamp around DNA, helping the DNA polymerase maintain contact with its template and thereby assisting with processivity. The inner face of the clamp enables DNA to be threaded through it. Once the polymerase reaches the end of the template or detects double stranded DNA, the sliding clamp undergoes a conformational change which releases the DNA polymerase. Clamp-loading proteins are used to initially load the clamp, recognizing the junction between template and RNA primers.

Eukaryotes

Within eukaryotes, DNA replication is controlled within the context of the cell cycle. As the cell grows and divides, it progresses through stages in the cell cycle; DNA replication occurs during the S phase (Synthesis phase).

The progress of the eukaryotic cell through the cycle is controlled by cell cycle checkpoints. Progression through checkpoints is controlled through complex interactions between various proteins, including cyclins and cyclin-dependent kinases. The G1/S checkpoint (or restriction checkpoint) regulates whether eukaryotic cells enter the process of DNA replication and subsequent division. Cells which do not proceed through this checkpoint are quiescent in the "G0" stage and do not replicate their DNA. Replication of chloroplast and mitochondrial genomes occurs independent of the cell cycle, through the process of D-loop replication.

Bacteria

Most bacteria do not go through a well-defined cell cycle and instead continuously copy their DNA; during rapid growth this can result in multiple rounds of replication occurring concurrently. Within *E coli,* the most well-characterized bacteria, regulation of DNA replication can be achieved through several mechanisms, including: the hemimethylation and sequestering of the origin sequence, the ratio of ATP to ADP, and the levels of protein DnaA. These all control the process of initiator proteins binding to the origin sequences.

Because *E coli* methylates GATC DNA sequences, DNA synthesis results in hemimethylated sequences. This hemimethylated DNA is recognized by a protein (SeqA) which binds and sequesters the origin sequence; in addition, dnaA (required for initiation of replication) binds less well to hemimethylated DNA. As a result, newly replicated origins are prevented from immediately initiating another round of DNA replication. ATP builds up when the cell is in a rich medium, triggering DNA replication once the cell has reached a specific size. ATP competes with ADP to bind to DnaA, and the DNA-ATP complex is able to initiate replication. A certain number of DnaA proteins are also required for DNA replication — each time the origin is copied the number of binding sites for DnaA doubles, requiring the synthesis of more DnaA to enable another initiation of replication.

TERMINATION OF REPLICATION

Because bacteria have circular chromosomes, termination of replication occurs when the two replication forks meet each other on the opposite end of the parental chromosome. *E coli* regulate this process through the use of termination sequences which, when bound by the Tus protein, enable only one direction of replication fork to pass through. As a result, the replication forks are constrained to always meet within the termination region of the chromosome.

Eukaryotes initiate DNA replication at multiple points in the chromosome, so replication forks meet and terminate at many points in the chromosome; these are not known to be regulated in any particular manner.

Because eukaryotes have linear chromosomes, DNA replication often fails to synthesize to the very end of the chromosomes (telomeres), resulting in telomere shortening. This is a normal process in somatic cells — cells are only able to divide a certain number of times before the DNA loss prevents further division. (This is known as the Hayflick limit.) Within the germ cell line, which passes DNA to the next generation, the enzyme telomerase extends the repetitive sequences of the telomere region to prevent degradation. Telomerase can become mistakenly active in somatic cells, sometimes leading to cancer formation.

ROLLING CIRCLE REPLICATION

Another method of copying DNA, sometimes used *in vivo* by bacteria and viruses, is the process of rolling circle replication. In this form of replication, a single replication fork progresses around a circular molecule to form multiple linear copies of the DNA sequence. In cells, this process can be used to rapidly synthesize multiple copies of plasmids or viral genomes.

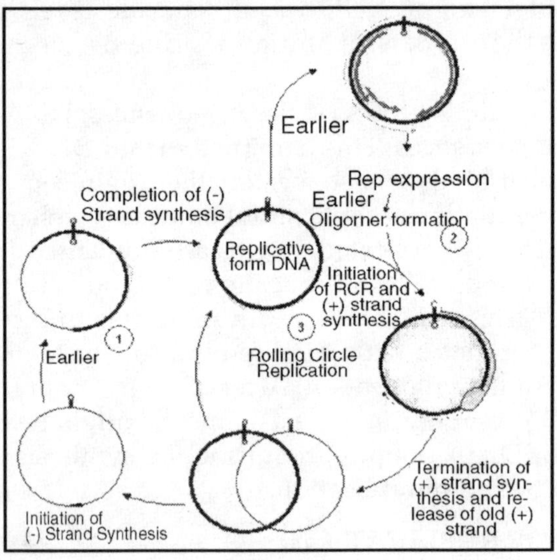

Fig. Rolling Circle Replication

In the cell, rolling circle replication is initiated by an initiator protein encoded by the plasmid or virus DNA. This protein is able to nick one strand of the double-stranded, circular DNA molecule at a site called the double-strand origin (DSO) and remains bound to the 5' phosphate end of the nicked strand. The free 3' hydroxyl end is released and can serve as a primer for DNA synthesis. Using the unnicked strand as a template, replication proceeds around the circular DNA molecule, displacing the nicked strand as single-stranded DNA. Continued DNA synthesis produces multiple single-stranded linear copies of the original DNA in a continuous head-to-tail series. *In vivo* these linear copies are subsequently converted to double-stranded circular molecules. Rolling circle replication can also be performed *in vitro* and has found wide uses in academic research and biotechnology, often used for amplification of DNA from very small amounts of starting material. Replication can be initiated by nicking a double-stranded circular DNA molecule or by hybridizing a primer to a single-stranded circle of DNA. The use of a reverse primer (or random primers) produces hyperbranched rolling circle amplification, resulting in exponential rather than linear growth of the DNA molecule.

POLYMERASE CHAIN REACTION

In vitro, researchers commonly replicate DNA using the polymerase chain reaction (PCR). PCR uses a pair of primers to span a target region in template DNA, polymerizing partner strands in each direction.

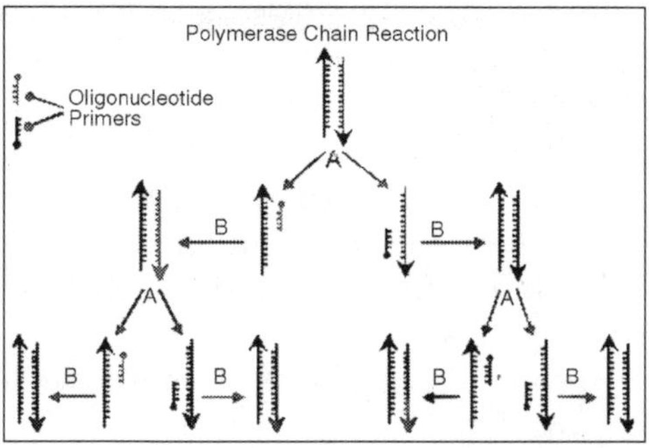

Fig. Polymerase Chain Reaction

This process can be repeated through multiple cycles through the use of a thermostable polymerase. At the start of each cycle, the mixture of template and primers is heated, separating the newly synthesized molecule and template. Then, as the mixture cools, both of these become templates for new primers to anneal to, and the polymerase extends from these. As a result the number of copies of the target region doubles each round, growing exponentially.

DNA IS A DOUBLE HELIX

The outer edges are formed of alternating ribose sugar molecules and phosphate groups. The two strands go in opposite directions (1 "up" and 1 "down"). The nitrogenous bases are "inside" like rungs on a ladder. Adenine on one side pairs with thymine (uracil in RNA) on the other by hydrogen bonding, and cytosine pairs with guanine. Note that the C-G pair has three hydrogen bonds while the A-T pair has only two, which keeps them from pairing wrong. This dictates side-to-side pairing, but says nothing about the order *along* the molecule. Watson and Crick said this variability along the molecule can account for the variety in the genetic code. Their model also accounts for how DNA can replicate itself. They said the molecule "unzips" and new matching bases are added in to create two new molecules. They called this semiconservative replication because each new molecule has one "old" and one "new" strand of DNA. DNA codes for protein synthesis by first coding for RNA. First, the DNA code is transcribed to RNA code, which is still in the "language" of nitrogenous bases, except that adenine on the DNA pairs with uracil (in place of thymine) on the RNA.

The RNA code is then translated to protein code, which is a different "language." This process involves ribosomes and two kinds of RNA: mRNA and tRNA. The mRNA codes for the gene in question and is copied off the DNA, while tRNA matches a specific group of nucleotides with a specific amino acid. A "unit" of three nucleotides on the tRNA codes for one amino acid.

Each of these "units" is called an anticodon. These match up with corresponding three-nucleotide sequences on the mRNA called codons, and in this manner the amino acids are organized into the correct sequence to build a protein. The ribosome works with the mRNA and tRNA to hook the amino acids together to form a protein. Here is a list of the mRNA codons and the corresponding amino acids for which they code. Mutations can be caused by a change in the sequence of the nucleotides. Some mutations have more effect than others, depending on where in the code they are and how important that area is to the code.

While mutations in some areas of some genes have little effect, sickle cell anemia is caused by a mutation in only one nucleotide. This changes the codon at that location to code for a different amino acid, and that, in turn, significantly changes the shape of the hemoglobin molecules in that person's blood. When some viruses (especially *Herpes* viruses, including Chicken Pox and Cold Sores) infect us, they insert their DNA into our cells' DNA, and stay resident in our cells for the rest of our lives. These can potentially become active again either making a person sick again or just being shed from a person's body (to infect others) without obvious symptoms of illness (like Mononucleosis).

Some kinds of cancer may be caused this way. For example, there is some pretty strong evidence linking genital warts (human papillomavirus, HPV) and cervical cancer. The AIDS virus does things "backwards." This virus contains RNA rather than DNA, yet when it gets into someone's cells, it can do reverse transcription and code from its RNA to make DNA which, then, can code to make more virus.

GENETIC ENGINEERING

We now have the knowledge and ability to transfer genes from one organism to another, which seems to have some benefits associated with it, but may also have many yet-to-be-discovered problems associated with it. Because this is all so new, not enough time has elapsed to allow scientists to study/look for any possible long-term effects of genetically-modified organisms (GMOs). Many medicines are now made by GMOs. For example, insulin was formerly extracted from the pancreas of animals after they were slaughtered for meat.

However, now most insulin is produced by bacteria with the insulin gene spliced onto their chromosome. Using this method of production, drug companies can make more insulin, faster. In theory, insulin produced in this way should be more "pure" — someone who could not use pig insulin due to a pork allergy may be able to tolerate insulin made in this way (but someone could, potentially, be allergic to some component of the bacteria present in the refined insulin).

In the relatively short time that this form of insulin has been available, there is no doubt that it has saved many people's lives — let's hope that some unforseen, long-term, deleterious effects are not discovered later on. Experimentation is being done to investigate the possible use of genetically-engineered viruses to treat genetic diseases such as cystic fibrosis. In this "treatment," a kind of virus that infects our lungs is used. The genes that enable it to infect our cells are kept while the genes which make us sick are (hopefully) all removed, and the missing human (normal) gene that relieves cystic fibrosis is then inserted into the virus' genome. These viruses are then sprayed into the lungs of a person with CF and allowed to "infect" the cells in that person's lungs.

When the normal gene is inserted into the genetic make-up of the cells lining that person's lungs, those cells function normally and the CF symptoms are alleviated. However, this treatment doesn't last because only that layer of cells is "infected," and when those cells die and are replaced by new cells, the new cells do not contain the genetic code to overcome the CF gene, and the person must inhale more genetically-engineered virus. While this seems to be a promising, life-giving, technique, no data are yet available on long-term effects and safety.

BACILLUS THURINGIENSIS (BT)

Bacillus thuringiensis (BT) is a species of bacterium that infects and kills a number of species of caterpillars (Remember that caterpillars turn into butterflies or moths when they grow up.). There is a species of moth whose caterpillar is a "pest" in corn plants, and insects of any kind are more successful when we humans plant huge monocultures of their favourite foods. For a number of years, now, people have realised that in certain situations, by judiciously applying BT, "pest" species of caterpillars can be infected and killed without the use of man-made, chemical insecticides.

More recently, a major US chemical company came up with the idea of creating genetically-engineered corn containing BT genes, which was good news to agri-business firms who plant huge areas of land with monocultures of corn. Because corn is wind-pollinated and therefore makes lots of pollen which, in this case, contains BT genes, many scientists are concerned about the effects of this corn on local butterfly populations. Some research has indicated that when this pollen settles on nearby caterpillar host plants and is, therefore, consumed by caterpillars, this might cause an increase in mortality (therefore fewer "good" butterflies such as Monarchs).

It is assumed that these BT genes in corn should, they think, have no adverse effects on people or cattle who consume this corn, but no long-term testing has been done. Also, this company has convinced the government that this corn should be marketed without any labeling indicating that it is genetically-engineered — they're scared that if we are given a choice, we won't buy/eat their corn if we know it is genetically engineered. Additionally, to increase their profits, this company has also put "suicide" genes into this corn so that farmers cannot save seed from one year to plant the next year, and have to buy more seed from them, instead.

For big agri-businesses, this is of small consequence, but for small, family farmers this is total disaster. To save money, the latter often save seed from one year to plant the next, and even if they don't plant this genetically-engineered corn, if their corn is pollinated by genetically-engineered corn from a neighboring field, it will not produce viable seed. This same chemical company came up with the idea of "transplanting" what they think is the cold-tolerance gene from a species of cold-water-inhabiting fish into tomatoes, thereby hoping to "invent" cold-tolerant tomatoes, and again, has convinced the government that these tomatoes should not be labeled in any way to indicate that they have fish genes in them, and that we should not have a choice about what we eat.

At the very least, this would be a problem for someone who is a vegetarian and chooses to not eat fish. A more serious consequence, however, would be that a person who is severely allergic to fish could also have an allergic reaction to these tomatoes and end up in the hospital. Again, no tests were done before the government approved these tomatoes and no data are available on the long-term effects on humans (or anything else).

This same chemical company manufactures a widely-used herbicide, and came up with the idea to genetically engineer cotton, soybeans, and other crop plants so they would be immune to the effects of that herbicide. That way, farmers can buy their seed from that company, then spray their fields with herbicide also purchased from that company to (hopefully) kill all local plants except the immune crop, simultaneously putting all their money in the corporation's pockets.

Preliminary research has shown that these resistant genes have already begun to "jump" into local weeds, thereby making them resistant to that herbicide, and again, no data are available on long-term effects on humans, other animals, or the environment in general.

STRUCTURE OF DNA

Most biological experiments are done with samples of living matter- cells, tissues, whole organisms or extracts of these materials. Only a few experiments are done with mathematical or physical models of biological components. Ocassionally, however, a model experiment gives an insight that would be difficult to obtain in any other way. This was true of the discovery of the structure of DNA. A crucial experiment was done by James Watson in 1952 using nothing more complicated than chemical models cut out of cardboard.

INHERITANCE

While the period from the early 1900s to World War II has been considered the "golden age" of genetics, scientists still had not determined that DNA, and not protein, was the hereditary material. However, during this time a great many genetic discoveries were made and the link between genetics and evolution was made. Friedrich Meischer in 1869 isolated DNA from fish sperm and the pus of open wounds. Since it came from nuclei, Meischer named this new chemical, nuclein. Subsequently the name was changed to nucleic acid and lastly to deoxyribonucleic acid (DNA).

Robert Feulgen, in 1914, discovered that fuchsin dye stained DNA. DNA was then found in the nucleus of all eukaryotic cells. During the 1920s, biochemist P.A. Levene analysed the components of the DNA molecule. He found it contained four nitrogenous bases: cytosine, thymine, adenine, and guanine; deoxyribose sugar; and a phosphate group. He concluded that the basic unit (nucleotide) was composed of a base attached to a sugar and that the phosphate also attached to the sugar. He (unfortunately) also erroneously concluded that the proportions of bases were equal and that there was a tetranucleotide that was the repeating structure of the molecule.

The nucleotide, however, remains as the fundemantal unit (monomer) of the nucleic acid polymer. There are four nucleotides: those with cytosine (C), those with guanine (G), those with adenine (A), and those with thymine (T). During the early 1900s, the study of genetics began in earnest: the link between Mendel's work and that of cell biologists resulted in the chromosomal theory of inheritance; Garrod proposed the link between genes and "inborn errors of metabolism"; and the question was formed: what is a gene?

The answer came from the study of a deadly infectious disease: *pneumonia*. During the 1920s Frederick Griffith studied the difference between a disease-causing strain of the pneumonia causing bacteria (*Streptococcus peumoniae*) and a strain that did not cause pneumonia. The pneumonia-causing strain (the S strain) was surrounded by a capsule. The other strain (the R strain) did not have a capsule and also did not cause pneumonia. Frederick Griffith was able to induce a nonpathogenic strain of the bacterium *Streptococcus pneumoniae* to become pathogenic. Griffith referred to a transforming factor that caused the non-pathogenic bacteria to become pathogenic. Griffith injected the different strains of bacteria into mice. The S strain killed the mice; the R strain did not. He further noted that if heat killed S strain was injected into a mouse, it did not cause pneumonia.

When he combined heat-killed S with Live R and injected the mixture into a mouse (remember neither alone will kill the mouse) that the mouse developed pneumonia and died. Bacteria recovered from the mouse had a capsule and killed other mice when injected into them!

Hypotheses:

- The dead S strain had been reanimated/resurrected.
- The Live R had been transformed into Live S by some "transforming factor".

Further experiments led Griffith to conclude that number 2 was correct. In 1944, Oswald Avery, Colin MacLeod, and Maclyn McCarty revisited Griffith's experiment and concluded the transforming factor was DNA. Their evidence was strong but not totally conclusive. The then-current favourite for the hereditary material was protein; DNA was not considered by many scientists to be a strong candidate.

The breakthrough in the quest to determine the hereditary material came from the work of Max Delbruck and Salvador Luria in the 1940s. Bacteriophage are a type of virus that attacks bacteria, the viruses that Delbruck and Luria worked with were those attacking *Escherichia coli,* a bacterium found in human intestines. Bacteriophages consist of protein coats covering DNA. Bacteriophages infect a cell by injecting DNA into the host cell. This viral DNA then "disappears" while taking over the bacterial machinery and beginning to make new virus instead of new bacteria. After 25 minutes the host cell bursts, releasing hundreds of new bacteriophage.

Phages have DNA and protein, making them ideal to resolve the nature of the hereditary material. In 1952, Alfred D. Hershey and Martha Chase conducted a series of experiments to determine whether protein or DNA was the hereditary material. By labeling the DNA and protein with different (and mutually exclusive) radioisotopes, they would be able to determine which chemical (DNA or protein) was getting into the bacteria.

Such material must be the hereditary material (Griffith's transforming agent). Since DNA contains Phosphorous (P) but no Sulpher (S), they tagged the DNA with radioactive Phosphorous-32. Conversely, protein lacks P but does have S, thus it could be tagged with radioactive Sulpher-35. Hershey and Chase found that the radioactive S remained outside the cell while the radioactive P was found inside the cell, indicating that DNA was the physical carrier of heredity.

Erwin Chargaff analysed the nitrogenous bases in many different forms of life, concluding that the amount of purines does not always equal the amount of pyrimidines (as proposed by Levene). DNA had been proven as the genetic material by the Hershey-Chase experiments, but how DNA served as genes was not yet certain. DNA must carry information from parent cell to daughter cell. It must contain information for replicating itself. It must be chemically stable, relatively unchanging. However, it must be capable of mutational change. Without mutations there would be no process of evolution.

Many scientists were interested in deciphering the structure of DNA, among them were Francis Crick, James Watson, Rosalind Franklin, and Maurice Wilkens. Watson and Crick gathered all available data in an attempt to develop a model of DNA structure. Franklin took X-ray diffraction photomicrographs of crystalline DNA extract, the key to the puzzle. The data known at the time was that DNA was a long molecule, proteins were helically coiled (as determined by the work of Linus Pauling), Chargaff's base data, and the x-ray diffraction data of Franklin and Wilkens.

DNA is a double helix, with bases to the centre (like rungs on a ladder) and sugar-phosphate units along the sides of the helix (like the sides of a twisted ladder). The strands are complementary (deduced by Watson and Crick from Chargaff's data, A pairs with T and C pairs with G, the pairs held together by hydrogen bonds). Notice that a double-ringed purine is always bonded to a single ring pyrimidine. Purines are Adenine (A) and Guanine (G). We have encountered Adenosine triphosphate (ATP) before, although in that case the sugar was ribose, whereas in DNA it is deoxyribose. Pyrimidines are Cytosine (C) and Thymine (T). The bases are complementary, with A on one side of the molecule we only get T on the other side, similarly with G and C. If we know the base sequence of one strand we know its complement.

HEREDITARY MATERIAL

DNA was proven as the hereditary material and Watson et al. had deciphered its structure. What remained was to determine how DNA copied its information and how that was expressed in the phenotype. Matthew Meselson and Franklin W. Stahl designed an experiment to determine the method of DNA replication. Three models of replication were considered likely.

Dispersive Replication

Involved the breaking of the parental strands during replication, and somehow, a reassembly of molecules that were a mix of old and new fragments on each strand of DNA. The Meselson-Stahl experiment involved the growth of *E. coli* bacteria on a growth medium containing heavy nitrogen (Nitrogen-15 as opposed to the more common, but lighter molecular weight isotope, Nitrogen-14). The first generation of bacteria was grown on a medium where the sole source of N was Nitrogen-15. The bacteria were then transferred to a medium with light (Nitrogen-14) medium. Watson and Crick had predicted that DNA replication was semi-conservative. If it was, then the DNA produced by bacteria grown on light medium would be intermediate between heavy and light. It was. DNA replication involves a great many building blocks, enzymes and a great deal of ATP energy (remember that after the S phase of the cell cycle cells have a G phase to regenerate energy for cell division). Only occurring in a cell once per (cell) generation, DNA replication in humans occurs at a rate of 50 nucleotides per second, 500/second in prokaryotes. Nucleotides have to be assembled and available in the nucleus, along with energy to make bonds between nucleotides.

DNA polymerases unzip the helix by breaking the H-bonds between bases. Once the polymerases have opened the molecule, an area known as the replication bubble forms (always initiated at a certain set of nucleotides, the origin of replication). New nucleotides are placed in the fork and link to the corresponding parental nucleotide already there (A with T, C with G). Prokaryotes open a single replication bubble, while eukaryotes have multiple bubbles. The entire length of the DNA molecule is replicated as the bubbles meet. Since the DNA strands are antiparallel, and replication proceeds in thje 5' to 3' direction on EACH strand, one strand will form a continuous copy, while the other will form a series of short Okazaki fragments.

ENZYME STRUCTURE OF DNA

Before a cell can begin to divide or differentiate, the genetic information within the cell's DNA must be copied, or "transcribed," onto complementary strands of RNA. RNA polymerase II (pol II) is an enzyme that, by itself, can unwind the DNA double helix, synthesize RNA, and proofread the result. When combined with other molecules that regulate and control the transcription process, pol II is the key to successful interpretion of an organism's genetic code.

However, the size, complexity, scarcity, and fragility of pol II complexes have made analysis of these macromolecules by x-ray crystallography a formidable challenge. A team of structural biologists has met this challenge using data obtained from both the Stanford Synchrotron Radiation Laboratory and the Macromolecular Crystallography Facility at the ALS. The resultant high-resolution model of a 10-subunit pol II complex suggests roles for each of the subunits and will allow researchers to begin unraveling the intricacies of DNA transcription and its role in gene expression.

In this work, the researchers studied the pol II enzyme from the yeast *Saccharomyces cerevisiae,* which is likely to be an excellent model for the human enzyme in light of its highly similar gene sequences. It is also the best-characterized form of the pol II enzyme, having been the subject of many biochemical and low-resolution structural studies in the past. To obtain a high-resolution structure, the research team drew on its considerable expertise in the preparation of protein crystals: two-dimensional crystals of pol II (minus two small subunits found to impede crystal growth) were used as seeds for growing three-dimensional crystals. These crystals, when produced in an inert atmosphere to prevent oxidation, enabled the collection of data to 3.5-angstrom resolution. The addition of a final soaking procedure to produce uniform crystals, combined with high-brightness x-ray sources, resulted in a resolution of 3.0 angstroms. The current results bring into focus the somewhat fuzzy features previously observed in or inferred from earlier experiments.

More importantly, the structural details suggest possible explanations for some of the unusual characteristics of this enzyme, which include a high processivity (the ability to synthesize very long strands of RNA) and the tendency to work in periodic spurts separated by pauses. While it is known that additional proteins (transcription factors) play a role in controlling the activity of pol II, scientists have yet to understand how such proteins interact with pol II binding sites to perform their various functions. The pol II model reported here establishes the positions of the various subunits and provides detailed information about the DNA/RNA binding domains.

The data reveal two main subunits separated by a deep cleft where DNA can enter the complex. At the end of the cleft is the active site, where the DNA can be unwound for a short distance (the "transcription bubble") and a DNA/RNA hybrid can be produced. Two prominent grooves lead away from the active site, either of which could accommodate the exiting RNA transcript. An opening below the active site may allow the entry of nucleotides (for manufacturing RNA) and transcription factors (for regulating the process). The same opening may provide room for the leading end of the RNA strand during "backtracking" maneuvers, which are important for proofreading and for traversing obstacles such as DNA damage.

Other notable features that might help account for the great stability of this transcribing complex include a pair of "jaws" that appear to grip the DNA strands as they enter the complex and, closer to the active site, a clamp on the DNA that could possibly be locked in the closed position by the presence of RNA. The high-resolution pol II structure reported here is a landmark achievement, pulling together threads from numerous diverse research efforts into a cohesive whole. Further study should yield many new insights into the detailed mechanisms of pol II and its transcription factors. Construction of an atomic model is already well underway.

UNRAVELING DNA

Encoded into the double-helical strands of DNA, the human genome is the complete set of instructions required to make a human being. While the mapping of the human genome (roughly three billion components) is certainly a Herculean accomplishment, it is only the first step towards realizing the full potential of genomic medicine in the diagnosis, monitoring, and treatment of disease. Beyond knowing what the genetic blueprint says, scientists must understand how that blueprint gets interpreted, or "expressed" as an individual with unique traits. The pol II enzyme is the catalyst for a major step in this process. As a pol II molecule slides along a DNA molecule, it "unzips" the strands of the DNA double helix, synthesizes a complementary strand of RNA (which will carry the genetic information to where it is needed), and verifies that no mistakes have occurred.

This process is regulated by transcription factors—separate molecules that bind to pol II and determine which genes are expressed, at what stage of development, and in which tissue. Done correctly, this process results in healthy cell growth and differentiation; otherwise, aberrations such as cancer can be the result. Thus, details of the structure of pol II, including information about its binding sites and how they interact with transcription factors, will provide valuable insight into the detailed mechanisms underlying the flow of genetic information from DNA to RNA to protein, which is necessary for life and health.

Without hydrogen bonds there could be no life because they hold the double helix of DNA together, and this they do by charge attractions. A hydrogen bonded to oxygen, or nitrogen, becomes slightly positively charged which enables it to attract a centre of negative charge on another molecule, such as another oxygen or nitrogen atom. The hydrogen bond is then written, *e.g.*, O–H…N, with the dotted line signifying the hydrogen bond. There are also O–H…O, N–H…O and N–H…N bonds, the last being among the weakest. The secondary effects they have on structures, molecular vibrations, etc., can be used to infer hydrogen bonding, but there is no primary way of observing them because of their inherent weakness. NMR appears to be the least useful technique because neither of the common isotopes, oxygen-16 or nitrogen-14, has a magnetic nucleus. However, nitrogen-15 has a magnetic moment, and by replacing ^{14}N by N, Grzesiek has opened up a new area of investigating these enigmatic bonds.

Paper reports for the first time the direct observation by NMR of an N–H…N hydrogen bond between nucleic acids enriched with ^{15}N, by measuring the coupling of the nitrogen atoms. Grzesiek, working with Andrew Dingley of the Heinrich-Heine University in Düsseldorf, has been able to do this and show that the coupling is surprisingly large. Normally atomic nuclei only couple with each other if they are linked by normal chemical bonds, and in theory hydrogen bonds have neither the strength nor stability for this to occur. The German researchers studied an ^{15}N enriched sample of the T1 domain of the potato spindle tuber viroid and were able to prove that N–H...N hydrogen bonding was present between the base pairs, uridine…adenosine and guanosine…cytidine, with couplings of approximately 7 Hz. How could they be certain that the signals they were observing are due to N–H…N hydrogen bonds?

The answer was to use triple resonance techniques to examine base pairs that hydrogen bond only via O–H…N hydrogen bonds, and show that the signal they had previously observed was absent. Grzesiek's second paper on hydrogen bonds, coauthored by Florence Cordier, Heinrich-Heine University, extends the work in an even more remarkable way by measuring the NMR coupling between nitrogens and carbons in the backbone hydrogen bonds of the human protein ubiquitin. The carbons are part of a carbonyl (C=O) group, so are one removed from the hydrogen bond, *i.e.*, N–H...O=C. This time they used material enriched with ^{15}N and ^{13}C (normal ^{12}C has no nuclear magnet), and there too was the evidence for these hydrogen bonds, albeit with an interaction an order of magnitude weaker (at –0.25 to –0.9 Hz) than the N–H...N coupling. Nevertheless, the couplings correlate with the strength of the hydrogen bond, being stronger in the stronger bonds.

Grzesiek's third paper, was done in conjunction with Dingley and researchers at UCLA. Together they studied not only the hydrogen bonding of Watson-Crick base pairs but also of Hoogsten base pairs within a DNA triplex consisting of one purine and two pyrimidine strands. Four different base pairs were identified, their various couplings distinguished–including those of the weaker interactions at the "frayed ends" of the DNA chains–and relationships with other hydrogen bonding parameters, such as bond length, were established. In addition they were able to show that density functional computer simulations by computer could reproduce these findings exactly.

DUPLICATION OF DNA

PROKARYOTIC DNA REPLICATION

Duplication of DNA is one of the fundamental properties of "molecule of life". Duplication or replication takes place once in a cell cycle. The duration and initiation point differs from one system to another. When conditions are favorable cell cytoplasmic mass increases and when cytoplasmic mass reaches a ratio to that of the cell size to 2L, bacterial cell division is triggered. To complete its cell cycle it requires hardly 30 -40 minutes. Several genes involved in different steps of cell division have been characterized. Par excellent system to understand molecular process of prokaryotic cell division is E.coli. The size of the E.coli chromosome is 4.6×10^6 base pairs, but it is compacted by the binding of histone like proteins. First the nucleoid i.e the chromosome frees from the membrane and unwinds. Such DNA initiates replication and completes in a matter of 15-20 minutes.

Then the cytoplasm starts dividing almost in the centre of the cell by central septal ring formation. It is at this time the daughter DNA molecules are partitioned and segregated to the two compartments. The central periseptal ring that started 1-2 minutes after DNA replication now furrows inwards and divides the cytoplasm in the middle region. Then mid-wall splits in the middle and daughter cell are set free, hence the Schizomycetes to bacteria (E.coli) Bacterial cell division requires the participation of the following genes and gene products. Eukaryotic cell division is more complex and highly regulated. The paradigm for it is yeast, for it is a unicellular system, grows fast and requires hardly 60 minutes for one cell cycle. One can obtain different kinds of mutation like conditional and null or auxotrophic kinds. In culture cells, the duration of one cell cycle is about 12 hrs. Onion root tip cells divide once in 12 hrs.

Whereas embryonic cells in its early stages require just 30 minutes for one cell cycle and they go through 6 to 8 such cell divisions very fast. Replication of DNA in cell division is an important phase. The replication is very accurate and it is nucleotide by nucleotide and there is no provision for making any mistakes, even if mistakes are made they are immediately corrected or repaired, otherwise consequences are serious and deleterious. In general DNA replication is precise, exact and regulated and involves initiation, elongation and termination steps, but mechanisms and rate of replication vary from one system to the other.

REPLICATION IS SEMI CONSERVATIVE

Meselson and Stahl showed that DNA replicates in semi conservative mode. There were some doubts about the mode whether it is conservative, dispersive or semi conservative. But Meselson and Stahl used nonradioactive heavy Nitrogen isotope (^{15}N), as the source of Nitrogen. When cells were grown on such source, ^{15}N incorporated nitrogen bases get incorporated into DNA, thus it becomes heavier than the DNA that is grown in normal (^{14}N) nitrogen media.

When two such DNAs, one grown in the presence of ^{15}N source and the other in ^{14}N (normal), are extracted separately and stained with ethidium Bromide, then subjected to equilibrium density centrifugation at ultra speed of 42000rpm for about 24 –36 hrs, the DNA, separates as two distinct bands, the one that is heavier (^{15}N) is the lower band than the normal (^{14}N) upper DNA. If cells grown in heavy isotopes are shifted to normal Nitrogen source and grown for one or two cell generation and if DNA is isolated from it and if one combine this with the first two DNA sources and subject them to equilibrium density ultra centrifugation (using cesium chloride or cesium sulphate), one can observe three bands, the top DNA band is from cells grown in ^{14}N without isotope, the second band is the hybrid band with one strand ^{15}N and the other with normal 14N-Nitrogen and the third band is with heavy nitrogen where both the strands are labeled with ^{15}N.

This experiment virtually and unambiguously proved that DNA replicates in semi conservative mode. Taylor using root tips, labeled with ^{32}P radioisotopes, of Vicia faba demonstrated semi conservative mode of replication at chromosomal level. Semi conservative means that the two daughter molecules produced at the end of replication, each of the molecules retains one parental strand and the other one is the newly made one. So it is semi conservative.

The mode of replication is absolutely semi conservative irrespective the kind of DNA, whether it is ds DNA or ssDNA, whether it is linear or circular, and the basic mechanism is more or less similar; each of them uses their own specialized mechanisms. Experimentally semiconservative mode of replication has been shown by density gradient centrifugation of non-radioactive isotopes such as ^{15}N and ^{14}N sources.

DNA SYNTHESIS

DNA Replication (also known as DNA synthesis) is a process where the double stranded Deoxyribonucleic Acid (DNA) is copied. Replication is an important first step in cell division, as cells must duplicate their entire genetic constitution before they can divide into two daughter cells. DNA replication is also a critical requirement for DNA repair.

REQUIREMENTS OF DNA REPLICATION

Replication requires three important components in order for this process to work.

Template Strand

The DNA serves as a template to guide the incoming nucleotides. The template strand attaches to the RNA primer strand in order to replicate the strand of DNA. The template just guides the nucleotides to the place that they need to go in. This is another component that is needed in order for DNA replication to even take place. If the DNA only had the template strand and none of the rest of the things then the strand could not replicate. The template starnd is very thin. Through an enzyme called primase the DNA template strand is synthesized one nucleotide at a time.

DNA Polymerase

There are many DNA polymerases within a cell, but only one of which is used in the replication of DNA. In order for DNA to replicate, it must have a primer, which is a small strand of RNA. The DNA template strand synthesizes one nucleotide at a time by the use of primase. The adding of nucleotides to the 3′ end, which is continued until that section of DNA is replicated. The other DNA polymerases that are within the cell are used in other ways besides replication. DNA polymerase III is one certain polymerase, which aids in replication of what is known as the "lagging strand". DNA ligase is the final enzyme, which combines the lagging strand

Free 3' Hydroxyl

The free 3' hydroxyl is the starter strand called a primer which is required in order for DNA to be replicated. The primer strand is not very big, it is shorter then a single strand of RNA. The primer is the complementary strand to the template strand of DNA. This is another one of the things that are needed in order to make the process of DNA replication even possible. The DNA polymerase is added to the 3' end of the primer, which keeps growing until the section is finally complete. After a while the RNA primer is degraded and disappears.

Types of Replication

During the discovery of DNA replication they also discovered that there are three modes of replication that a DNA strand can take. The three types of DNA replication are semiconservative, conservative, and dispersive. These three types of replication were tested on in order to see the possible patterns that would result in the complementary base pairing.

Semiconservative Replication

Through this type of replication the parent strand serves as a template for the new strand. The two offsprings would have one of the parent strand and one new strand.

Conservative Replication

Through this type of replication the double helix serves as a template. This although does not contribute to the new double helix.

Dispersive Replication

In this type of replication the fragments from the parent DNA molecule. The DNA serves as a template for the assembly of the new molecule. The new double helix contains old and new parts of the DNA strands.

DNA Repairing

After the DNA is replicated, there are three DNA repair mechanisms.

They are to be used when they feel necessary. This will lessen the chance of human mutation. The rate of error in which the DNA polymerases makes a mistake is very high. This is why the DNA has mechanisms to cause fewer mutations. The three mechanisms are:

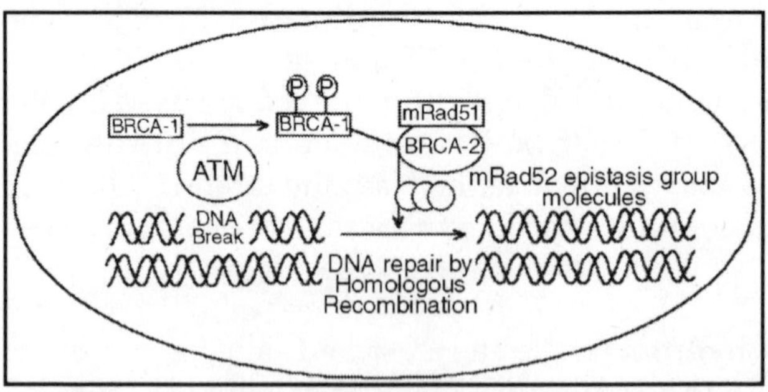

Fig. DNA Repairing

Proofreading

Every time a nucleotide is introduced to an already growing chain, the DNA polymerase checks the connection. If the pair is mismatched, it will be removed and redone. This lowers the overall rate of mutations.

Mismatch Pair

After the replication takes place, another set of proteins check to make sure that there are no more mismatched base pairs and also which strand is the wrong one.

Excision Pair

There can still be damage while the cell is living. Because of this, there are enzymes constantly checking the cell. If they detect a problem, the enzyme cuts the strand and also cuts away the opposite base. DNA polymerase and ligase then fix this base sequence.

LABORATORY TECHNIQUE

Scientists use two techniques in a laboratory that involve DNA replication. They use these techniques to observe genes and genomes. One of the techniques can make multiple copies of DNA from only a short piece. Another technique allows scientists to determine the base sequence of DNA. The two techniques are:

Polymerase Chain Reaction (PCR)

Through this process, a short strand of DNA is copied repetitively. There are three steps that are repeated in the process. The DNA strands are denatured or heated in order to separate the strand. Next, they add a primer, which was artificially made.

They also add DNA polymerase and the four deoxyribon-ucleotide triphosphates. These three are needed in order to replicate DNA. The final step is then the DNA polymerase catalyzes the production of a complementary strand.

DNA Sequencing

This allows scientists to determines the base sequence of a DNA. The process is the DNA is first denatured and a single strand is placed in a test tube. They then add DNA polymerase, primer, the four dNTP's, and small amounts of ddNTP's.

The DNA replicates within the tube and after the DNA fragments are denatured. The fragments then go through electrophoresis, which sorts the lengths of DNA fragments. The fragments then pass through a laser beam which excites the fluorescent tags. This information is fed into a computer at which tells the sequence.

Mendelson-Stahl Experiment

Through this experiment Meselson- Stahl convinced the scientists that the correct model for DNA replication was the semiconservative replication. They used density labeling in order to distinguish the old and new strands of DNA. In their experiment they used "heavy" isotopes of nitrogen. This nitrogen is a nonradioactive isotope that made the molecule more dense. The chemically identical molecules containing an isotope known as ^{14}N. They needed a way to distinguish the different DNA densities.

Because of this they needed to come up with something to measure the densities of solutions. Meselson, Stahl, and Jerome Vanguard came up with centrifuge, which is a procedure that involves the spinning of solutions at high speed, which causes the particles to separate and form gradient according to the different densities of the solution. The first part of their experiment they grew cultures of Escherichia coli. One of the cultures was grown in ^{15}N, which made the DNA "heavy".

The other culture was placed in a medium using ^{14}N rather than ^{15}N, which made all the DNA "light". In order to see if the use of centrifuge worked they combined the two cultures and when centrifuged it formed two DNA bands. This showed them the different densities with in the two cultures. The next part of their experiment they grew E. Coli in the ^{15}N medium and then transferred the bacteria from that medium to the ^{14}N. Through this they came up with the conclusion that through E. Coli DNA replicates every 20 minutes. They then took DNA samples from each generation through, which they found that the density gradient was different each generation.

Their observations could only be explained with the semiconservative model of DNA replication. Their results showed that when the DNNA first underwent replication, it was in 15N, which caused the DNA to be heavy. Because in the semiconservative model of DNA replication, the one strand acts as a template for a second strand which the DNA was in a ^{14}N DNA strand and a ^{15}N and they were of intermediate density.

THE ORDERED STEPS IN THE CELL CYCLE

G1: The first gap in the normal cell cycle is called G1 and is the period when the necessary proteins for DNA replication are synthesized. However, this phase of the cell cycle is not only characterized by synthesis of replication machinery.

During this period the cell must monitor both the internal and external environments to ensure that all the preparations for DNA synthesis have been completed and that overall conditions for cell division are favorable. As discussed below, there is a major check-point in a normal cell cycle that is critical for ensuring that all is well for the cell to enter S-phase.

S-phase: The duplication of the cellular content of DNA occurs during S-phase, so-called because this is the phase when DNA is synthesized. This phase of the cell cycle is the longest taking 10-12 hours of a typical 24hr eukaryotic cell cycle.

G2: During the second gap phase of the cell cycle the cell undertakes the synthesis of the proteins required to assemble the machinery required for separation of the duplicated chromosomes (the process called mitosis) and ultimately division of the parental cell into two daughter cells (the process termed cytokinesis).

Like the G1 phase, the G2 phase is also a stage when the internal and external environments are monitored to ensure that faithful replication of the DNA has occurred and that conditions are favorable for cytokinesis. In addition, as for the G1 phase there is a major check-point at the end of the G2 phase that controls the entry into M-phase.

M-phase: During M-phase there is an ordered series of events that leads to the alignment and separation of the duplicated chromosomes (called sister chromatids) This process is divided into distinct steps that were originally identified and characterized through light microscopic observations of dividing cells.

The steps of mitosis are termed prophase, prometaphase, metaphase, anaphase and telophase. Although cytokinesis is the process by which the parental cell is physically separated into two new daughter cells, it actually begins during anaphase. The processes that occur during M-phase require much less time than those of S-phase, generally lasting only 1-2 hrs.

During prophase the duplicated chromosomes condense while outside the nucleus the mitotic spindle assembles between the two centrosomes. The centrosome is an organelle that serves as the main microtubule organizing centre that is involved in the attachment of microtubules to the sister chromatids.

During prometaphase the nuclear membrane breaks apart and the chromosomes can attach to spindle microtubules and begin active movement. During metaphase the chromosomes are aligned at the equator of the spindle midway between the spindle poles. The sister chromatids are attached to opposite poles of the spindle.

During anaphase the sister chromatids synchronously separate to form the two sets of daughter chromosomes. Each sister chromatid is slowly pulled towards the spindle pole it faces. During telophase the two daughter chromosomes arrive at the spindle poles and decondense. A new nuclear envelope forms around each set of chromosomes which forms the two new nuclei. This process marks the end of mitosis and sets the stage for cytokinesis.

CHECKPOINTS AND CELL CYCLE REGULATION

It should seem obvious that the processes that drive a cell through the cell cycle must be highly regulated so as to ensure that the resultant daughter cells are viable and each contains the complement of DNA found in the original parental cell. There are many "parts" to the systems that control the transit through a eukaryotic cell cycle. These "parts" include mechanisms to control the timing of events so that each individual process is turned on and off at the appropriate time, mechanisms to initiate each event in the correct order and to also ensure that each event is triggered only once per cell cycle, controls to ensure events occur in a linear, irreversible direction, redundancy, or back-ups to ensure the cycle functions properly even in the context of some malfunctioning parts, and systems that are adaptable so that cell cycle events can be modified in the context of different cell types and/or environmental conditions.

Many of the most important discoveries about the mechanisms that control events of the cell cycle were elucidated using yeasts which are single cell eukaryotes. By analysis of various mutants that inactivated genes encoding essential components of cell cycle control systems in yeast many important control genes were identified. These genes were identified as cell division cycle genes or cdc genes. Thus, many cell cycle control genes in mammalian cells are also called cdc genes. Much of the control of the progression through the phases of a cell cycle are exerted at checkpoints. There are many such checkpoints but the two most critical are those that occur near the end of G1 prior to S-phase entry and those near the end of G2 prior to mitosis.

As indicated above, there is the need for cell cycle control mechanisms to exert their influences at specific times during each transit through a cell cycle. The heart of this timing control is the responsibility of a family of protein kinases that are called cyclin-dependent kinases, CDKs. The kinase activity of these enzymes rises and falls as the cell progresses through a cell cycle. Different CDKs operate at different points in the cell cycle. As would be expected, the oscillating changes in the activity of CDKs leads to oscillating changes in the phosphorylation of various intracellular proteins.

These phosphorylations alter the activity of the modified proteins which then effect changes in events of the cell cycle. The cyclical activity of each CDK is controlled by a complex series of proteins, the most important of which are the cyclins, hence the name of the enzymes as cyclin-dependent kinases. The CDKs are absolutely dependent upon their interaction with the cyclins for activity.

Unless they are tightly bound CDKs have no kinase activity. The cyclins were originally idenitified because they undergo a cycle of synthesis and degradation at specific points in each cell cycle. Thus, whereas the levels of the various CDKs remain fairly constant throughout the cell cycle, their activities changes in concert with the fluctuations of the cyclins. Four different classes of cyclins have been defined on the basis of the stage of the cell cycle in which they bind and activate CDKs. These four classes are G1-cyclins, G1/S-cyclins, S-cyclins, and M-cyclins. The cyclin nomenclature and associated CDK in mammalian cells are listed in the following Table.

Cyclin-CDK Complex	Cyclin	CDK Partner
G1-CDK	cyclin D*	CDK4, CDK6
G1/S-CDK	cyclin E	CDK2
S-CDK	cyclin A	CDK2
M-CDK	cyclin B	CDK1**

- There are three D cyclins in mammals: D1, D2, and D3
- CDK1 is the same as CDC2 in fission yeast and CDC28 in budding yeast

The G1-cyclins are not found in all eukaryotic cells but in those where they are synthesized they promote passage through a restriction point in late G1 called Start. The G1/S-cyclins bind to their cognate CDKs at the end of G1 and it is this interaction that is required to commit the cell to the process of DNA replication in S-phase. The S-cyclins bind to their cognate CDKs during S-phase and it is this interaction that is required for the initiation of DNA synthesis. The M-cyclins bind to their cognate CDKs and in so doing promote the events of mitosis.

Although CDKs are inactive unless bound to a cyclin, there is more to the activation process than just the interaction of the two parts of the complex. When cyclins bind to CDKs they alter the conformation of the CDK resulting in exposure of a domain that is the site of phosphorylation by another kinase called CDK-activating kinase (CAK). Following phosphorylation the cyclin-CDK complex is fully active.

In addition to control of CDK kinase activity by cyclin binding and CAK phosphorylation, control is exerted to inhibit CDK activity through interaction with inhibitory proteins as well as by inhibitory phosphorylation events. Thus, there is extremely tight control on the overall activity of each CDK. One of the inhibitory kinases that phosphorylates CDKs is called Wee1. The inhibitory phosphorylations are removed through the action of a phosphatase called CDC25.

The action of these two regulatory enzymes on CDK activity is most important at the level of the M-CDK activity at the onset of mitosis. Proteins that bind to and inhibit cyclin-CDK complexes are called CDK inhibitory proteins (CKI, for cyclin-kinase inhibitor). Mammalian cells express two classes of CKI. These are called CIPs for CDK inhibitory proteins and INK4 for inhibitors of kinase 4. The CIPs bind and inhibit CDK1, CDK2, CDK4, and CDK6 complexes, whereas the INK4s bind and inhibit only the CDK4 and CDK6 complexes.

There are at least three CIP proteins in mammalian cells and these are identified as $p^{21Cip1/WAF1}$ (gene symbol=CDKN1A), p^{27KIP1} (gene symbol=CDKN1B), and p^{57KIP2} (gene symbol=CDKN1C). The expression of each of these CIPs is controlled by specific events that may have occurred during cell cycle transit. For example p^{21Cip1} expression is induced in response to DNA damage. This induction is under the control of the action of the tumor suppressor protein p53. There are at least four INK4 proteins that are each identified by their molecular weights: $p^{15INK4B}$, $p^{16INK4A}$, and $p^{18INK4C}$ (these were the first 3 characterized) as well as p^{19INK4}. The $p^{16INK4A}$ protein is also a tumor suppressor since loss of its function leads to cancer. All the INK4 proteins contain 4 tandem repeats of a sequence of amino acids that were first identified in ankyrin and are thus referred to as ankyrin repeats.

As indicated above, many cells reside in a resting or quiescent state but can be stimulated by external signals to re-enter the cell cycle. These external growth promoting signals are the result of growth factors binding to their receptors. Most growth factors induce the expression of genes that are referred to as early and delayed-response genes. The activation of early response genes occurs in response to growth factor receptor-mediated signal transduction resulting in phosphorylation and activation of transcription factor proteins that are already present in the cell.

Many of the induced early response genes are themselves transcription factors that in turn activate the expression of delayed-response genes. In the context of the cell cycle, these delayed-response genes encode proteins of the G1-CDK complexes. One such early response gene is the proto-oncogene MYC. With respect to the cell cycle some of the genes turned on by activation of MYC are cyclin D, proteins of the ubiquitin ligase complex called SCF (Skp1/cullin/F-box protein) and the members of the E2F transcription factor family.

There are six members of the E2F family: E2F1 through E2F6). The synthesis of cyclin D will result in the activation of G1-CDK complexes. The synthesis of components of SCF leads to the degradation of p27KIP1 which normally inhibits G1-CDK complexes. The synthesis of E2F family members results in increased synthesis of proteins involved in DNA synthesis as well as the synthesis of the S-phase cyclins A and E and CDK2. Regulation of E2F activity by the tumor suppressor pRB will be discussed below. The cyclical degradation of the cyclins is effected through the action of several different ubiquitin ligase complexes. The action of ubiquitin ligases in protein turnover is discussed in more detail in the Protein Modifications page.

There are two important ubiquitin ligase complexes that control the turnover of cyclins and other cell cycle regulating proteins. One is the SCF complex which functions to control the transit from G1 to S-phase and the other is called anaphase promoting complex (APC) which controls the levels of the M-phase cyclins as well as other regulators of mitosis.

One important function of APC is to control the initiation of sister chromatid separation which begins at the metaphase-anaphase transition. The attachment of the sister chromatids to the opposite poles of the mitotic spindles occurs early during mitosis. The ability of the sister chromatids to be pulled apart is initially inhibited because they are bound together by a protein complex termed cohesin complex.

The cohesin complex is deposited along the chromosomes as they are duplicated during S-phase. Anaphase can only begin with the disruption of the cohesin complex. The breakdown of the cohesin complex is brought about as a consequence of the activation of the ubiquitin ligase activity of the APC. APC targets a protein called securin. Securin functions to inhibit the protease called separase and the action of separase is to degrade the proteins of the cohesin complex, thus allowing sister chromatid separation.

ISOLATION OF GENOMIC DNA

PCR allows isolation of DNA fragments from genomic DNA by selective amplification of a specific region of DNA. This use of PCR augments many methods, such as generating hybridization probes for Southern or northern hybridization and DNA cloning, which require larger amounts of DNA, representing a specific DNA region. PCR supplies these techniques with high amounts of pure DNA, enabling analysis of DNA samples even from very small amounts of starting material.

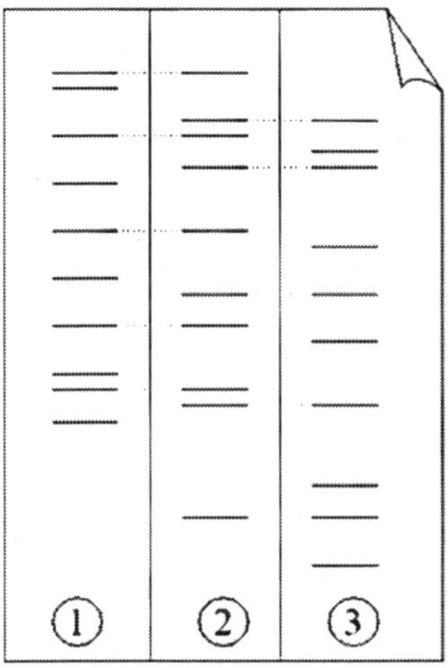

Fig: Electrophoresis of PCR-Amplified DNA Fragments. (1) Father. (2) Child. (3) Mother. The Child has Inherited some, but not all of the Fingerprint of Each of its Parents, Giving it a New, Unique Fingerprint.

Other applications of PCR include DNA sequencing to determine unknown PCR-amplified sequences in which one of the amplification primers may be used in Sanger sequencing, isolation of a DNA sequence to expedite recombinant DNA technologies involving the insertion of a DNA sequence into a plasmid or the genetic material of another organism.

Bacterial colonies (E.coli) can be rapidly screened by PCR for correct DNA vector constructs. PCR may also be used for genetic fingerprinting; a forensic technique used to identify a person or organism by comparing experimental DNAs through different PCR-based methods. Some PCR 'fingerprints' methods have high discriminative power and can be used to identify genetic relationships between individuals, such as parent-child or between siblings, and are used in paternity testing. This technique may also be used to determine evolutionary relationships among organisms.

Amplification and Quantitation of DNA

Because PCR amplifies the regions of DNA that it targets, PCR can be used to analyse extremely small amounts of sample. This is often critical for forensic analysis, when only a trace amount of DNA is available as evidence. PCR may also be used in the analysis of ancient DNA that is thousands of years old. These PCR-based techniques have been successfully used on animals, such as a forty-thousand-year-old mammoth, and also on human DNA, in applications ranging from the analysis of Egyptian mummies to the identification of a Russian Tsar.

Quantitative PCR methods allow the estimation of the amount of a given sequence present in a sample – a technique often applied to quantitatively determine levels of gene expression. Real-time PCR is an established tool for DNA quantification that measures the accumulation of DNA product after each round of PCR amplification.

PCR in Diagnosis of Diseases

PCR allows early diagnosis of malignant diseases such as leukemia and lymphomas, which is currently the highest developed in cancer research and is already being used routinely. PCR assays can be performed directly on genomic DNA samples to detect translocation-specific malignant cells at a sensitivity which is at least 10,000 fold higher than other methods.PCR also permits identification of non-cultivatable or slow-growing microorganisms such as mycobacteria, anaerobic bacteria, or viruses from tissue culture assays and animal models.

The basis for PCR diagnostic applications in microbiology is the detection of infectious agents and the discrimination of non-pathogenic from pathogenic strains by virtue of specific genes.Viral DNA can likewise be detected by PCR. The primers used need to be specific to the targeted sequences in the DNA of a virus, and the PCR can be used for diagnostic analyses or DNA sequencing of the viral genome. The high sensitivity of PCR permits virus detection soon after infection and even before the onset of disease. Such early detection may give physicians a significant lead in treatment. The amount of virus ("viral load") in a patient can also be quantified by PCR-based DNA quantitation techniques.

VARIATIONS ON THE BASIC PCR TECHNIQUE

- *Allele-specific PCR*: This diagnostic or cloning technique is used to identify or utilize single-nucleotide polymorphisms (SNPs) (single base differences in DNA). It requires prior knowledge of a DNA sequence, including differences between alleles, and uses primers whose 3' ends encompass the SNP. PCR amplification under stringent conditions is much less efficient in the presence of a mismatch between template and primer, so successful amplification with an SNP-specific primer signals presence of the specific SNP in a sequence.

- *Assembly PCR or Polymerase Cycling Assembly (PCA)*: Assembly PCR is the artificial synthesis of long DNA sequences by performing PCR on a pool of long oligonucleotides with short overlapping segments. The oligonucleotides alternate between sense and antisense directions, and the overlapping segments determine the order of the PCR fragments thereby selectively producing the final long DNA product.
- *Asymmetric PCR*: Asymmetric PCR is used to preferentially amplify one strand of the original DNA more than the other. It finds use in some types of sequencing and hybridization probing where having only one of the two complementary stands is required. PCR is carried out as usual, but with a great excess of the primers for the chosen strand. Due to the slow (arithmetic) amplification later in the reaction after the limiting primer has been used up, extra cycles of PCR are required. A recent modification on this process, known as Linear-After-The-Exponential-PCR (LATE-PCR), uses a limiting primer with a higher melting temperature (Melting temperature|Tm) than the excess primer to maintain reaction efficiency as the limiting primer concentration decreases mid-reaction.
- *Helicase-dependent amplification*: This technique is similar to traditional PCR, but uses a constant temperature rather than cycling through denaturation and annealing/extension cycles. DNA Helicase, an enzyme that unwinds DNA, is used in place of thermal denaturation.
- *Hot-start PCR*: This is a technique that reduces non-specific amplification during the initial set up stages of the PCR. The technique may be performed manually by heating the reaction components to the melting temperature (*e.g.*, 95 °C) before adding the polymerase. Specialized enzyme systems have been developed that inhibit the polymerase's activity at ambient temperature, either by the binding of an antibody or by the presence of covalently bound inhibitors that only dissociate after a high-temperature activation step. Hot-start/cold-finish PCR is achieved with new hybrid polymerases that are inactive at ambient temperature and are instantly activated at elongation temperature.
- *Intersequence-specific (ISSR) PCR*: A PCR method for DNA fingerprinting that amplifies regions between some simple sequence

repeats to produce a unique fingerprint of amplified fragment lengths.

- *Inverse PCR*: A method used to allow PCR when only one internal sequence is known. This is especially useful in identifying flanking sequences to various genomic inserts. This involves a series of DNA digestions and self ligation, resulting in known sequences at either end of the unknown sequence.
- *Ligation-mediated PCR*: This method uses small DNA linkers ligated to the DNA of interest and multiple primers annealing to the DNA linkers; it has been used for DNA sequencing, genome walking, and DNA footprinting.
- *Methylation-specific PCR (MSP)*: The MSP method was developed by Stephen Baylin and Jim Herman at the Johns Hopkins School of Medicine, and is used to detect methylation of CpG islands in genomic DNA. DNA is first treated with sodium bisulfite, which converts unmethylated cytosine bases to uracil, which is recognized by PCR primers as thymine. Two PCRs are then carried out on the modified DNA, using primer sets identical except at any CpG islands within the primer sequences. At these points, one primer set recognizes DNA with cytosines to amplify methylated DNA, and one set recognizes DNA with uracil or thymine to amplify unmethylated DNA. MSP using qPCR can also be performed to obtain quantitative rather than qualitative information about methylation.
- *Miniprimer PCR*: Miniprimer PCR uses a novel thermostable polymerase (S-Tbr) that can extend from short primers ("smalligos") as short as 9 or 10 nucleotides, instead of the approximately 20 nucleotides required by Taq. This method permits PCR targeting smaller primer binding regions, and is particularly useful to amplify unknown, but conserved, DNA sequences, such as the 16S (or eukaryotic 18S) rRNA gene. 16S rRNA miniprimer PCR was used to characterize a microbial mat community growing in an extreme environment, a hypersaline pond in Puerto Rico. In that study, deeply divergent sequences were discovered with high frequency and included representatives that deûned two new division-level taxa, suggesting that miniprimer PCR may reveal new dimensions of microbial diversity. By enlarging the "sequence space" that may be queried by PCR primers, this technique may enable novel PCR strategies that are not possible within the limits of primer design imposed by Taq and other commonly used enzymes.
- Multiplex Ligation-dependent Probe Amplification (MLPA): permits multiple targets to be amplified with only a single primer pair, thus avoiding the resolution limitations of multiplex PCR.

- Multiplex-PCR: The use of multiple, unique primer sets within a single PCR mixture to produce amplicons of varying sizes specific to different DNA sequences. By targeting multiple genes at once, additional information may be gained from a single test run that otherwise would require several times the reagents and more time to perform. Annealing temperatures for each of the primer sets must be optimized to work correctly within a single reaction, and amplicon sizes, *i.e.*, their base pair length, should be different enough to form distinct bands when visualized by gel electrophoresis.
- Nested PCR: increases the specificity of DNA amplification, by reducing background due to non-specific amplification of DNA. Two sets of primers are being used in two successive PCRs. In the first reaction, one pair of primers is used to generate DNA products, which besides the intended target, may still consist of non-specifically amplified DNA fragments. The product(s) are then used in a second PCR with a set of primers whose binding sites are completely or partially different from and located 3' of each of the primers used in the first reaction. Nested PCR is often more successful in specifically amplifying long DNA fragments than conventional PCR, but it requires more detailed knowledge of the target sequences.
- Overlap-extension PCR: is a genetic engineering technique allowing the construction of a DNA sequence with an alteration inserted beyond the limit of the longest practical primer length.
- *Quantitative PCR (Q-PCR)*: is used to measure the quantity of a PCR product (preferably real-time). It is the method of choice to quantitatively measure starting amounts of DNA, cDNA or RNA. Q-PCR is commonly used to determine whether a DNA sequence is present in a sample and the number of its copies in the sample. The method with currently the highest level of accuracy is Quantitative real-time PCR. It is often confusingly known as RT-PCR (Real Time PCR) or RQ-PCR. QRT-PCR or RTQ-PCR are more appropriate contractions. RT-PCR commonly refers to reverse transcription PCR, which is often used in conjunction with Q-PCR. QRT-PCR methods use fluorescent dyes, such as Sybr Green, or fluorophore-containing DNA probes, such as TaqMan, to measure the amount of amplified product in real time.
- *RT-PCR*: (Reverse Transcription PCR) is a method used to amplify, isolate or identify a known sequence from a cellular or tissue RNA. The PCR is preceded by a reaction using reverse transcriptase to convert RNA to cDNA. RT-PCR is widely used in expression profiling, to determine the expression of a gene or to identify the sequence of an RNA transcript, including transcription start and

termination sites and, if the genomic DNA sequence of a gene is known, to map the location of exons and introns in the gene. The 5' end of a gene (corresponding to the transcription start site) is typically identified by an RT-PCR method, named RACE-PCR, short for *Rapid Amplification of cDNA Ends*.

- *Solid Phase PCR*: Encompasses multiple meanings, including Polony Amplification (where PCR colonies are derived in a gel matrix, for example), 'Bridge PCR' (the only primers present are covalently linked to solid support surface), conventional Solid Phase PCR (where Asymmetric PCR is applied in the presence of solid support bearing primer with sequence matching one of the aqueous primers) and Enhanced Solid Phase PCR (where conventional Solid Phase PCR can be improved by employing high Tm solid support primer with application of a thermal 'step' to favour solid support priming).
- *TAIL-PCR*: Thermal asymmetric interlaced PCR is used to isolate unknown sequence flanking a known sequence. Within the known sequence TAIL-PCR uses a nested pair of primers with differing annealing temperatures; a degenerate primer is used to amplify in the other direction from the unknown sequence.
- *Touchdown PCR*: a variant of PCR that aims to reduce nonspecific background by gradually lowering the annealing temperature as PCR cycling progresses. The annealing temperature at the initial cycles is usually a few degrees (3-5ÚC) above the T_m of the primers used, while at the later cycles, it is a few degrees (3-5ÚC) below the primer T_m. The higher temperatures give greater specificity for primer binding, and the lower temperatures permit more efficient amplification from the specific products formed during the initial cycles.
- *PAN-AC*: This method uses isothermal conditions for amplification, and may be used in living cells.
- *Universal Fast Walking*: this method allows genome walking and genetic fingerprinting using a more specific 'two-sided' PCR than conventional 'one-sided' approaches (using only one gene-specific primer and one general primer-which can lead to artefactual 'noise') by virtue of a mechanism involving lariat structure formation. Streamlined derivatives of UFW are LaNe RAGE (lariat-dependent nested PCR for rapid amplification of genomic DNA ends), 5'RACE LaNe and 3'RACE LaNe.

HISTORY

A 1971 paper in the Journal of Molecular Biology by Kleppe and co-workers first described a method using an enzymatic assay to replicate a short DNA template with primers *in vitro*. However, this early manifestation of the basic PCR principle did not receive much attention, and the invention of the polymerase chain reaction in 1983 is generally credited to Kary Mullis. At the core of the PCR method is the use of a suitable DNA polymerase able to withstand the high temperatures of >90°C (>195°F) required for separation of the two DNA strands in the DNA double helix after each replication cycle. The DNA polymerases initially employed for in vitro experiments presaging PCR were unable to withstand these high temperatures. So the early procedures for DNA replication were very inefficient, time consuming, and required large amounts of DNA polymerase and continual handling throughout the process.

A 1976 discovery of Taq polymerase a DNA polymerase purified from the thermophilic bacterium, *Thermus aquaticus,* which naturally occurs in hot (50 to 80 °C (120 to 175 °F)) environments paved the way for dramatic improvements of the PCR method. The DNA polymerase isolated from *T. aquaticus* is stable at high temperatures remaining active even after DNA denaturation, thus obviating the need to add new DNA polymerase after each cycle.

This allowed an automated thermocycler-based process for DNA amplification. At the time he developed PCR in 1983, Mullis was working in Emeryville, California for Cetus Corporation, one of the first biotechnology companies. There, he was responsible for synthesizing short chains of DNA. Mullis has written that he conceived of PCR while cruising along the Pacific Coast Highway one night in his car. He was playing in his mind with a new way of analyzing changes (mutations) in DNA when he realised that he had instead invented a method of amplifying any DNA region through repeated cycles of duplication driven by DNA polymerase.

In *Scientific American,* Mullis summarized the procedure: "Beginning with a single molecule of the genetic material DNA, the PCR can generate 100 billion similar molecules in an afternoon. The reaction is easy to execute. It requires no more than a test tube, a few simple reagents, and a source of heat." He was awarded the Nobel Prize in Chemistry in 1993 for his invention, seven years after he and his colleagues at Cetus first put his proposal to practice. However, some controversies have remained about the intellectual and practical contributions of other scientists to Mullis' work, and whether he had been the sole inventor of the PCR principle.

Patent Wars

The PCR technique was patented by Cetus Corporation, where Mullis worked when he invented the technique in 1983. The *Taq* polymerase enzyme was also covered by patents. There have been several high-profile lawsuits related to the technique, including an unsuccessful lawsuit brought by DuPont. The pharmaceutical company Hoffmann-La Roche purchased the rights to the patents in 1992 and currently holds those that are still protected. A related patent battle over the Taq polymerase enzyme is still ongoing in several jurisdictions around the world between Roche and Promega. The legal arguments have extended beyond the life of the original PCR and Taq polymerase patents.

DNA FINGERPRINTING

DNA Fingerprinting, method of identification that compares fragments of deoxyribonucleic acid (DNA). It is sometimes called DNA typing. DNA is the genetic material found within the cell nuclei of all living things. In mammals the strands of DNA are grouped into structures called chromosomes. With the exception of identical twins, the complete DNA of each individual is unique.

A DNA fingerprint is constructed by first extracting a DNA sample from body tissue or fluid such as hair, blood, or saliva. The sample is then segmented using enzymes, and the segments are arranged by size using a process called electrophoresis. The segments are marked with probes and exposed on X-ray film, where they form a characteristic pattern of black bars—the DNA fingerprint. If the DNA fingerprints produced from two different samples match, the two samples probably came from the same person. DNA fingerprinting was first developed as an identification technique in 1985. Originally used to detect the presence of genetic diseases, DNA fingerprinting soon came to be used in criminal investigations and forensic science. The first criminal conviction based on DNA evidence in the United States occurred in 1988. In criminal investigations, DNA fingerprints derived from evidence collected at the crime scene are compared to the DNA fingerprints of suspects. The DNA evidence can implicate or exonerate a suspect. Generally, courts have accepted the reliability of DNA testing and admitted DNA test results into evidence.

However, DNA fingerprinting is controversial in a number of areas: the accuracy of the results, the cost of testing, and the possible misuse of the technique. The accuracy of DNA fingerprinting has been challenged for several reasons. First, because DNA segments rather than complete DNA strands are "fingerprinted," a DNA fingerprint may not be unique; large-scale research to confirm the uniqueness of DNA fingerprinting test results has not been conducted. In addition, DNA fingerprinting is often performed in private laboratories that may not follow uniform testing standards and quality controls. Also, since human beings must interpret the test, human error could lead to false results. DNA fingerprinting is expensive. Suspects who are unable to provide their own DNA experts may not be able to adequately defend themselves against charges based on DNA evidence.

In the United States, the Federal Bureau of Investigation (FBI) has created a national database of genetic information called the National DNA Index System. The database contains DNA obtained from convicted criminals and from evidence found at crime scenes. Some experts fear that this database might be used for unauthorized purposes, such as identifying individuals with stigmatizing illnesses such as acquired immunodeficiency syndrome (AIDS).

DNA SEQUENCING

The term DNA sequencing encompasses biochemical methods for determining the order of the nucleotide bases, adenine, guanine, cytosine, and thymine, in a DNA oligonucleotide. The sequence of DNA constitutes the heritable genetic information in nuclei, plasmids, mitochondria, and chloroplasts that forms the basis for the developmental programmes of all living organisms.

Determining the DNA sequence is therefore useful in basic research studying fundamental biological processes, as well as in applied fields such as diagnostic or forensic research. The advent of DNA sequencing has significantly accelerated biological research and discovery. The rapid speed of sequencing attained with modern DNA sequencing technology has been instrumental in the large-scale sequencing of the human genome, in the Human Genome Project. Related projects, often by scientific collaboration across continents, have generated the complete DNA sequences of many animal, plant, and microbial genomes.

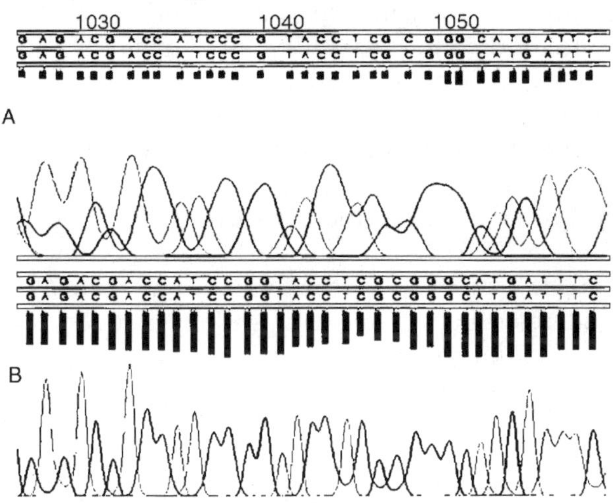

Fig. DNA Sequence Trace

Early Methods

For thirty years, a large proportion of DNA sequencing has been carried out with the chain-termination method developed by Frederick Sanger and coworkers in 1975. Prior to the development of rapid DNA sequencing methods in the early 1970s by Sanger in England and Walter Gilbert and Allan Maxam at Harvard, a number of laborious methods were used. For instance, in 1973 Gilbert and Maxam reported the sequence of 24 basepairs using a method known as wandering-spot analysis. RNA sequencing, which for technical reasons is easier to perform than DNA sequencing, was one of the earliest forms of nucleotide sequencing.

The major landmark of RNA sequencing, dating from the pre-recombinant DNA era, is the sequence of the first complete gene and then the complete genome of Bacteriophage MS2..

Maxam-Gilbert Sequencing

In 1976-1977, Allan Maxam and Walter Gilbert developed a DNA sequencing method based on chemical modification of DNA and subsequent cleavage at specific bases.

Chemical sequencing method two years after the ground-breaking paper of Sanger and Coulson on plus-minus sequencing, Maxam-Gilbert sequencing rapidly became more popular, since purified DNA could be used directly, while the initial Sanger method required that each read start be cloned for production of single-stranded DNA. However, with the development and improvement of the chain-termination method Maxam-Gilbert sequencing has fallen out of favour due to its technical complexity, extensive use of hazardous chemicals, and difficulties with scale-up.

In addition, unlike the chain-termination method, chemicals used in the Maxam-Gilbert method cannot easily be customised for use in a standard molecular biology kit. In brief, the method requires radioactive labelling at one end and purification of the DNA fragment to be sequenced. Chemical treatment generates breaks at a small proportion of one or two of the four nucleotide bases in each of four reactions (G, A+G, C, C+T).

Thus a series of labelled fragments is generated, from the radiolabelled end to the first 'cut' site in each molecule. The fragments are then size-separated by gel electrophoresis, with the four reactions arranged side by side. To visualize the fragments generated in each reaction, the gel is exposed to X-ray film for autoradiography, yielding an image of a series of dark 'bands' corresponding to the radiolabelled DNA fragments, from which the sequence may be inferred. Also sometimes known as 'chemical sequencing', this method originated in the study of DNA-protein interactions (footprinting), nucleic acid structure and epigenetic modifications to DNA, and within these it still has important applications.

CHAIN-TERMINATION METHODS

While the chemical sequencing method of Maxam and Gilbert, and the plus-minus method of Sanger and Coulson were orders of magnitude faster than previous methods, the chain-terminator method developed by Sanger was even more efficient, and rapidly became the method of choice. The Maxam-Gilbert technique requires the use of highly toxic chemicals, and large amounts of radiolabeled DNA, whereas the chain-terminator method uses fewer toxic chemicals and lower amounts of radioactivity. The key principle of the Sanger method was the use of dideoxynucleotides triphosphates (ddNTPs) as DNA chain terminators.

The classical chain-termination or Sanger method requires a single-stranded DNA template, a DNA primer, a DNA polymerase, radioactively or fluorescently labeled nucleotides, and modified nucleotides that terminate DNA strand elongation. The DNA sample is divided into four separate sequencing reactions, containing the four standard deoxynucleotides (dATP, dGTP, dCTP and dTTP) and the DNA polymerase.

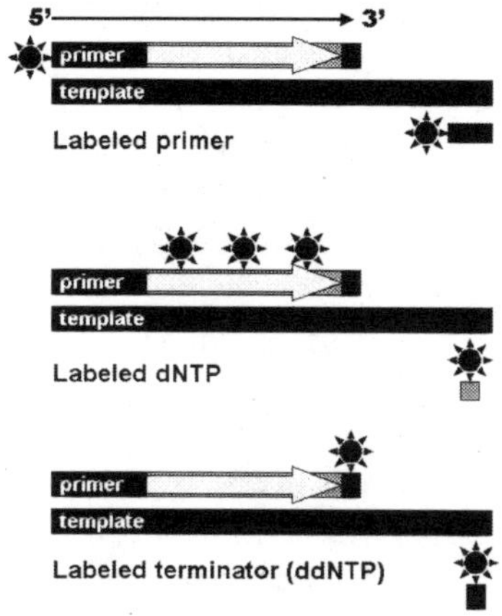

Fig: Part of a radioactively labelled sequencing gel

To each reaction is added only one of the four dideoxynucleotides (ddATP, ddGTP, ddCTP, or ddTTP). These dideoxynucleotides are the chain-terminating nucleotides, lacking a 3'-OH group required for the formation of a phosphodiester bond between two nucleotides during DNA strand elongation. Incorporation of a dideoxynucleotide into the nascent (elongating) DNA strand therefore terminates DNA strand extension, resulting in various DNA fragments of varying length. The dideoxynucleotides are added at lower concentration than the standard deoxynucleotides to allow strand elongation sufficient for sequence analysis.

The newly synthesized and labeled DNA fragments are heat denatured, and separated by size (with a resolution of just one nucleotide) by gel electrophoresis on a denaturing polyacrylamide-urea gel. Each of the four DNA synthesis reactions is run in one of four individual lanes (lanes A, T, G, C); the DNA bands are then visualized by autoradiography or UV light, and the DNA sequence can be directly read off the X-ray film or gel image. In the image on the right, X-ray film was exposed to the gel, and the dark bands correspond to DNA fragments of different lengths. A dark band in a lane indicates a DNA fragment that is the result of chain termination after incorporation of a dideoxynucleotide (ddATP, ddGTP, ddCTP, or ddTTP).

The terminal nucleotide base can be identified according to which dideoxynucleotide was added in the reaction giving that band. The relative positions of the different bands among the four lanes are then used to read (from bottom to top) the DNA sequence as indicated.There are some technical variations of chain-termination sequencing. In one method, the DNA fragments are tagged with nucleotides containing radioactive phosphorus for radiolabelling.

Alternatively, a primer labeled at the 5′ end with a fluorescent dye is used for the tagging. Four separate reactions are still required, but DNA fragments with dye labels can be read using an optical system, facilitating faster and more economical analysis and automation. This approach is known as 'dye-primer sequencing'. The later development by L Hood and coworkers of fluorescently labeled ddNTPs and primers set the stage for automated, high-throughput DNA sequencing. The different chain-termination methods have greatly simplified the amount of work and planning needed for DNA sequencing.

For example, the chain-termination-based "Sequenase" kit from USB Biochemicals contains most of the reagents needed for sequencing, prealiquoted and ready to use. Some sequencing problems can occur with the Sanger method, such as non-specific binding of the primer to the DNA, affecting accurate read-out of the DNA sequence. In addition, secondary structures within the DNA template, or contaminating RNA randomly priming at the DNA template can also affect the fidelity of the obtained sequence. Other contaminants affecting the reaction may consist of extraneous DNA or inhibitors of the DNA polymerase.

Dye-Terminator Sequencing

An alternative to primer labelling is labelling of the chain terminators, a method commonly called 'dye-terminator sequencing'. The major advantage of this method is that the sequencing can be performed in a single reaction, rather than four reactions as in the labelled-primer method.

In dye-terminator sequencing, each of the four dideoxynucleotide chain terminators is labelled with a different fluorescent dye, each fluorescing at a different wavelength. This method is attractive because of its greater expediency and speed and is now the mainstay in automated sequencing with computer-controlled sequence analysers. Its potential limitations include dye effects due to differences in the incorporation of the dye-labelled chain terminators into the DNA fragment, resulting in unequal peak heights and shapes in the electronic DNA sequence trace chromatogram after capillary electrophoresis.

This problem has largely been overcome with the introduction of new DNA polymerase enzyme systems and dyes that minimize incorporation variability, as well as methods for eliminating "dye blobs", caused by certain chemical characteristics of the dyes that can result in artifacts in DNA sequence traces. The dye-terminator sequencing method, along with automated high-throughput DNA sequence analysers, is now being used for the vast majority of sequencing projects, as it is both easier to perform and lower in cost than most previous sequencing methods.

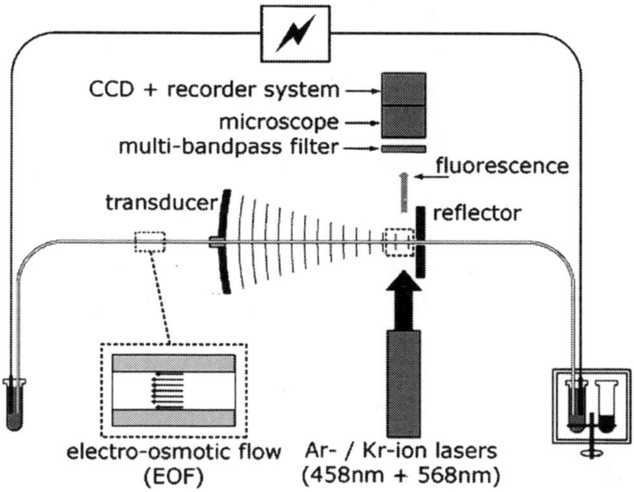

Fig: Capillary Electrophoresis

Challenges

Modern sequencing typically produces a sequence that has poor quality in the first 15-40 bases, a high quality region of 700-900 bases, and then quickly deteriorating quality. Base calling software typically outputs an estimate of quality along with the sequence to aid in quality trimming. Before the DNA can be sequenced, linker sequences are attached to its ends, and it is inserted into a cloning vector. The resulting sequence can therefore often contain parts of the vector or the linker sequences, which must be filtered out prior to analysis. In contrast, emerging sequencing technologies based on pyrosequencing often avoid using cloning vectors.

During PCR amplification, unrelated sequences can hybridize, and the resulting clone can be a chimaeric sequence, containing fragments from both sequences. Another problem is polymerase stuttering, where the polymerase repeatedly outputs the same fragments, giving an artificially long low-complexity part of the sequence.

Automation and Sample Preparation

Modern automated DNA sequencing instruments (DNA sequencers) can sequence up to 384 fluorescently labelled samples in a single batch (run) and perform as many as 24 runs a day. However, automated DNA sequencers carry out only DNA-size-based separation (by capillary electrophoresis), detection and recording of dye fluorescence, and data output as fluorescent peak trace chromatograms. Sequencing reactions by thermocycling, cleanup and re-suspension in a buffer solution before loading onto the sequencer are performed separately.

In the past, an operator had to trim the low quality ends of every sequence manually in order to remove the sequencing errors. However, today, software like Fast Chromatogram Viewer can automatically trim the ends at batch.

LARGE-SCALE SEQUENCING STRATEGIES

Current methods can directly sequence only relatively short (300-1000 nucleotides long) DNA fragments in a single reaction. The main obstacle to sequencing DNA fragments above this size limit is insufficient power of separation for resolving large DNA fragments that differ in length by only one nucleotide. Limitations on ddNTP incorporation were largely solved by Tabor at Harvard Medical, Carl Fuller at USB biochemicals, and their coworkers.Large-scale sequencing aims at sequencing very long DNA fragments. Even relatively small bacterial genomes contain millions of nucleotides, and the human chromosome 1 alone contains about 246 million bases.

Therefore, some approaches consist of cutting (with restriction enzymes) or shearing (with mechanical forces) large DNA fragments into shorter DNA fragments. The fragmented DNA is cloned into a DNA vector, usually a bacterial artificial chromosome (BAC), and amplified in *Escherichia coli*. The amplified DNA can then be purified from the bacterial cells (a disadvantage of bacterial clones for sequencing is that some DNA sequences may be inherently *un-clonable* in some or all available bacterial strains, due to deleterious effect of the cloned sequence on the host bacterium or other effects). These short DNA fragments purified from individual bacterial colonies are then individually and completely sequenced and assembled electronically into one long, contiguous sequence by identifying 100%-identical overlapping sequences between them (shotgun sequencing).

This method does not require any pre-existing information about the sequence of the DNA and is often referred to as *de novo* sequencing. Gaps in the assembled sequence may be filled by Primer walking, often with sub-cloning steps (or transposon-based sequencing depending on the size of the remaining region to be sequenced). These strategies all involve taking many small *reads* of the DNA by one of the above methods and subsequently assembling them into a contiguous sequence.

The different strategies have different tradeoffs in speed and accuracy; the shotgun method is the most practical for sequencing large genomes, but its assembly process is complex and potentially error-prone-particularly in the presence of sequence repeats. Because of this, the assembly of the human genome is not literally complete — the repetitive sequences of the centromeres, telomeres, and some other parts of chromosomes result in gaps in the genome assembly. Despite having only 93% of the full genome assembled, the Human Genome Project was declared complete because their definition of human genome sequencing was limited to euchromatic sequence (99% complete at the time), excluding these intractable repetitive regions.

The human genome is about 3 billion (3,000,000,000) bp long; if the average fragment length is 500 bases, it would take a minimum of six million (3 billion/500) to sequence the human genome (not allowing for overlap = 1-fold coverage). Keeping track of such a high number of sequences presents significant challenges, only held down by developing and coordinating several procedural and computational algorithms, such as efficient database development and management.

Resequencing or *targeted sequencing* is utilized for determining a change in DNA sequence from a "reference" sequence. It is often performed using PCR to amplify the region of interest (pre-existing DNA sequence is required to design the PCR primers). Resequencing uses three steps, extraction of DNA or RNA from biological tissue; amplification of the RNA or DNA (often by PCR); followed by sequencing. The resultant sequence is compared to a reference or a normal sample to detect mutations.

NEW SEQUENCING METHODS

High-Throughput Sequencing

The high demand for low cost sequencing has given rise to a number of high-throughput sequencing technologies. These efforts have been funded by public and private institutions as well as privately researched and commercialized by biotechnology companies. High-throughput sequencing technologies are intended to lower the cost of sequencing DNA libraries beyond what is possible with the current dye-terminator method based on DNA separation by capillary electrophoresis. Many of the new high-throughput methods use methods that parallelize the sequencing process, producing thousands or millions of sequences at once.

As molecular detection methods are often not sensitive enough for single molecule sequencing, most approaches use an *in vitro* cloning step to generate many copies of each individual molecule. Emulsion PCR is one method, isolating individual DNA molecules along with primer-coated beads in aqueous bubbles within an oil phase.

A polymerase chain reaction (PCR) then coats each bead with clonal copies of the isolated library molecule and these beads are subsequently immobilized for later sequencing. (commercialized by 454 Life Sciences, acquired by Roche), Shendure and Porreca et al. (also known as "polony sequencing") and Solid sequencing, (developed by Agencourt and acquired by Applied Biosystems). Another method for *in vitro* clonal amplification is "bridge PCR", where fragments are amplified upon primers attached to a solid surface, developed and used by Solexa (now owned by Illumina). These methods both produce many physically isolated locations which each contain many copies of a single fragment. The single-molecule method developed by Stephen Quake's laboratory (later commercialized by Helicos) skips this amplification step, directly fixing DNA molecules to a surface.

6

Genetic Engineering

Genetically modified plants are created by the process of genetic engineering, which allows scientists to move genetic material between organisms with the aim of changing their characteristics. All organisms are composed of cells that contain the DNA molecule. Molecules of DNA form units of genetic information, known as genes. Each organism has a genetic blueprint made up of DNA that determines the regulatory functions of its cells and thus the characteristics that make it unique. Prior to genetic engineering, the exchange of DNA material was possible only between individual organisms of the same species.

With the advent of genetic engineering in 1972, scientists have been able to identify specific genes associated with desirable traits in one organism and transfer those genes across species boundaries into another organism. A gene from bacteria, virus, or animal may be transferred into plants to produce genetically modified plants having changed characteristics. Thus, this method allows mixing of the genetic material among species that cannot otherwise breed naturally. The success of a genetically improved plant depends on the ability to grow single modified cells into whole plants. Some plants like potato and tomato grow easily from single cell or plant tissue. Others such as corn, soy bean, and wheat are more difficult to grow. After years of research, plant specialists have been able to apply their knowledge of genetics to improve various crops such as corn, potato, and cotton. They have to be careful to ensure that the basic characteristics of these new plants are the same as the traditional ones, except for the addition of the improved traits.

The world of biotechnology has always moved fast, and now it is moving even faster. More traits are emerging; more land than ever before is being planted with genetically modified varieties of an ever-expanding number of crops. Research efforts are being made to genetically modify most plants with a high economic value such as cereals, fruits, vegetables, and floriculture and horticulture species.

BIOLOGICAL CELL

It only takes one biological cell to create an organism. In fact, there are countless species of single celled organisms, and indeed multi-cellular organisms like ourselves.

A single cell is able to keep itself functional by owning a series of *'miniature machines'* known as *organelles*. The following list looks at some of these organelles and other characteristics typical of a fully functioning cell.

- *Mitochondrion:* An important cell organelle involved in respiration
- *Cytoplasm:* A fluid surrounding the contents of a cell and forms a vacuole
- *Golgi Apparatus:* The processing area for the creation of a glycoprotein
- *Endoplasmic Reticulum:* An important organelle heavily involved in protein synthesis.
- *Vesicles: Packages* of substances that are to be used in the cell or secreted by it.
- *Nucleus*: The "brain" of a cell containing genetic information that determines every natural process within an organism.
- *Cell Membrane:* Also known as a plasma membrane, this outer layer of a cell assists in the movement of molecules in and out the cell plays both a structural and protective role
- *Lysosomes:* Membranous sacs that contain digestive enzymes

CELL WALL

A structure that characteristically is found in plants and prokaryotes and not animals that plays a structural and protective role.

CELL SPECIALISATION

Cells can become specialised to perform a particular function within an organism, usually as part of a larger tissue consisting of many of the same cells working in tandem:

- *Nerve cells* to operate as part of the nervous system to send messages back and forth via the brain at the centre of the nerve system.
- Skin cells for waterproof protection and protection against pathogens in the open air environment.
- *Xylem* tubes to transport water around plants and to provide structural support for the plant as a whole.

Cells combine their efforts in these tissue types to perform a common cause. The task of the specialised cell will determine in what way it is going to be specialised, because different cells are suited to different purposes, as illustrated in the above list:

- Muscle cells are long and smooth in structure and their elastic nature allows these cells to perform flexible movements, just as they do in our own body's.
- Some *white blood cells* contain powerful digestive enzymes to

eliminate pathogens by breaking them down to the molecular level.

- Cells at the back of the eye are sensitive to light stimuli, and thus can *interpret* differences in light intensity which can in turn be interpreted by our nervous system and brain.

Many of these cells contain organelles, though after some cells are specialised, they do not possess particular characteristics as they do not require them to be there. *i.e.* efficiency is the key, no resources are wasted and the resources available are put to their idyllic optimum.

THE CELL MEMBRANE

The *cell membrane,* otherwise known as the plasma membrane is a semi-permeable structure consisting mainly of *phospholipid* (fat) molecules and proteins. They are structured in a *fluid mosaic model,* where a double layer of phospholipid molecules provide a barrier accompanied by proteins. It is present round the circumference of a cell to acts as a barrier, keeping foreign entities out the cell and its contents (like cytoplasm) firmly inside the cell. The plasma membrane allows only selected materials to pass in and out of a cell, and is thus known as a selectively permeable membrane.

CELL TRANSPORT

There are three methods in which ions are transported through the cell membrane into the cell,

- *Active Transport*: Active transport is the transport of molecules with the active assistance of a carrier that can transport the material against a natural *concentration gradient.*
- *Passive Transport (Diffusion):* The movement of molecules from areas of high concentration (*i.e.* outside a cell) to areas of low concentration (*i.e.* within a cell) via a carrier. This process does not require energy.
- *Simple Diffusion:* The movement of molecules from areas of high concentration to areas of low concentration in a free state. *Osmosis* of water involves this type of diffusion through a selectively permeable membrane (*i.e.* plasma membrane)

THE BREAKDOWN OF MATERIALS IN A CELL

In cells, sometimes it is required to breakdown more complex molecules into more simple molecules, which can then be 're-built' into what is needed by the body with these new raw materials. '*Pinocytosis*' where to contents of a structure (such as bacteria) are *drank,* essentially by breaking down molecules into a drinkable form. '*Phagocytosis*' where contents are 'eaten'.

ABSORPTION AND SECRETION

Absorption is the uptake of materials from a cells' external environment. Secretion is the ejection of material.

PLANT GENETICS

Knowledge of population genetics, quantitative genetics, probability theory and statistics is indispensable for understanding equilibria and shifts with regard to the genotypic composition of a population, its mean value and its variation. The subject of population genetics is the study of equilibria and shifts of allele and genotype frequencies in populations.

These equilibria and shifts are determined by five forces:

- Mode of reproduction of the considered crop

The mode of reproduction is of utmost importance with regard to the breeding of any particular crop and the maintenance of already available varieties. This applies both to the natural mode of reproduction of the crop and to enforced modes of reproduction, like those applied when producing a hybrid variety. In plant breeding theory, crops are therefore classified into the following categories: cross-fertilizing crops, self-fertilizing crops, crops with both cross- and self-fertilization and asexually reproducing crops. It is explained that even within a specific population, traits may differ with regard to their mode of reproduction.

- Selection
- Mutation
- Immigration of plants or pollen, *i.e.* immigration of alleles
- Random variation of allele frequencies

A population is a group of (potentially) interbreeding plants occurring in a certain area, or a group of plants originating from one or more common ancestors. The former situation refers to cross-fertilizing crops (in which case the term Mendelian population is sometimes used), while the latter group concerns, in particular, self-fertilizing crops. In the absence of immigration the population is said to be a closed population.

Examples of closed populations are:

- A group of plants belonging to a cross-fertilizing crop, grown in an isolated field, *e.g.* maize or rye (both pollinated by wind), or turnips or Brussels sprouts (both pollinated by insects)
- A collection of lines of a self-fertilizing crop, which have a common origin, *e.g.* a single-cross, a three-way cross, a backcross

The subject of quantitative genetics concerns the study of the effects of alleles and genotypes and of their interaction with environmental conditions. Population genetics is usually concerned with the probability distribution of genotypes within a population (genotypic composition), while quantitative genetics considers phenotypic values (and statistical parameters dealing with them, especially mean and variance) for the trait under investigation.

In fact population genetics and quantitative genetics are applications of probability theory in genetics. An important subject is, consequently, the derivation of probability distributions of genotypes and the derivation of expected genotypic values and of variances of genotypic values. Generally, statistical analyses comprise estimation of parameters and hypothesis testing. In quantitative genetics statistics is applied in a number of ways. It begins when considering the experimental design to be used for comparing entries in the breeding programme.

Considered across the entries constituting a population (plants, clones, lines, families) the expression of an observed trait is a random variable. If the expression is represented by a numerical value the variable is generally termed phenotypic value, represented by the symbol p.

Two genetic causes for variation in the expression of a trait are distinguished. Variation controlled by so-called major genes, *i.e.* alleles that exert a readily traceable effect on the expression of the trait, is called qualitative variation. Variation controlled by so-called polygenes, *i.e.* alleles whose individual effects on a trait are small in comparison with the total variation, is called quantitative variation. In Note: it is elaborated that this classification does not perfectly coincide with the distinction between qualitative traits and quantitative traits.

The former paragraph suggests that the term *gene* and *allele* are synonyms. According to Rieger, Michaelis and Green a gene is a continuous region of DNA, corresponding to one (or more) transcription units and consisting of a particular sequence of nucleotides. Alternative forms of a particular gene are referred to as alleles. In this respect the two terms 'gene' and 'allele' are sometimes interchanged. Thus the term 'gene frequency' is often used instead of the term 'allele frequency'. The term locus refers to the site, alongside a chromosome, of the gene/allele.

Since the term 'gene' is often used as a synonym of the term 'locus', we have tried to avoid confusion by preferential use of the terms 'locus' and 'allele' (as a synonym of the word gene) where possible. In the case of qualitative variation, the phenotypic value p of an entry (plant, line, family) belonging to a genetically heterogeneous population is a discrete random variable. The phenotype is then exclusively (or to a largely traceable degree) a function f of the genotype, which is also a random variable G.

Thus,

$$p = f(G)$$

It is often desired to deduce the genotype from the phenotype. This is possible with greater or lesser correctness, depending for example on the degree of dominance and sometimes also on the effect of the growing conditions on the phenotype. A knowledge of population genetics suffices for an insight into the dynamics of the genotypic composition of a population with regard to a trait with qualitative variation: application of quantitative genetics is then superfluous.

Note: All traits can show both qualitative and quantitative variation. Culm length in cereals, for instance, is controlled by dwarfing genes with major effects, as well as by polygenes. The commonly used distinction between qualitative traits and quantitative traits is thus, strictly speaking, incorrect. When exclusively considering qualitative variation, *e.g.* with regard to the traits in pea (*Pisum sativum*) studied by Mendel, this book describes the involved trait as a trait showing qualitative variation. On the other hand, with regard to traits where quantitative variation dominates – and which are consequently mainly discussed in terms of this variation – one should realise that they can also show qualitative variation.

In this sense the following economically important traits are often considered to be 'quantitative characters':

- Biomass
- Yield with regard to a desired plant product
- Content of a desired chemical compound (oil, starch, sugar, protein, lysine) or an undesired compound
- Resistance, including components of partial resistance, against biotic or abiotic stress factors
- Plant height

In the case of quantitative variation p results from the interaction of a complex genotype, *i.e.* several to many loci are involved, and the specific growing conditions are important. In this book, by complex genotype we mean the sum of the genetic constitutions of all loci affecting the expression of the considered trait. These loci may comprise loci with minor genes (or polygenes), as well as loci with major genes, as well as loci with both. With regard to a trait showing quantitative variation, it is impossible to classify individual plants, belonging to a genetically heterogeneous population, according to their genotypes. This is due to the number of loci involved and the complicating effect on p of (some) variation in the quality of the growing conditions. It is, thus, impossible to determine the number of plants representing a specified complex genotype. (With regard to the expression of qualitative variation this may be possible!). Knowledge of both population genetics and quantitative genetics is therefore required for an insight into the inheritance of a trait with quantitative variation.

The phenotypic value for a quantitative trait is a continuous random variable and so one may write:

$$p = f(G, e)$$

Thus the phenotypic value is a function f of both the complex genotype (represented by G) and the quality of the growing conditions (say environment, represented by e). Even in the case of a genetically homogeneous group of plants (a clone, a pure line, a single-cross hybrid) p is a continuous random variable.

The genotype is a constant and one should then write:

$$p = f(G, e)$$

Regularly in this book, simplifying assumptions will be made when developing quantitative genetic theory.

Especially the following assumptions will often be made:

- Absence of linkage of the loci controlling the studied trait(s)
- Absence of epistatic effects of the loci involved in complex genotypes.

These assumptions will now be considered.

Absence of Linkage

The assumption of absence of linkage for the loci controlling the trait of interest, *i.e.* the assumption of independent segregation, may be questionable in specific cases, but as a generalisation it can be justified by the following reasoning. Suppose that each of the n chromosomes in the genome contains M loci affecting the considered trait. This implies presence of n groups of,

$$\binom{M}{2} -$$

pairs of loci consisting of loci which are more strongly or more weakly linked. The proportion of pairs consisting of linked loci among all pairs of loci amounts then to,

$$\frac{n\binom{M}{2}}{\binom{nM}{2}} = \frac{n.M!}{2!(M-2)!} \times \frac{2!(nM-2)!}{(nM)!} = \frac{M-1}{nM-1} = \frac{1-\frac{1}{M}}{n-\frac{1}{M}}$$

For $M = 1$ this proportion is 0; for $M = 2$ it amounts to 0.077 for rye (*Secale cereale,* with $n = 7$) and to 0.024 for wheat (*Triticum aestivum,* with $n = 21$); for $M = 3$ it amounts to 0.100 for rye and to 0.032 for wheat. For $M \to \infty$ the proportion is 1 n; *i.e.* 0.142 for rye and 0.048 for wheat.

One may suppose that loci located on the same chromosome, but on different sides of the centromere, behave as unlinked loci. If each of the n chromosomes contains,

$$m\left(=\frac{1}{2}M\right)$$

relevant loci on each of the two arms then there are $2n$ groups of,

$$\binom{m}{2} -$$

pairs consisting of linked loci. Thus considered, the proportion of pairs consisting of linked loci amounts to,

$$\frac{2n\binom{m}{2}}{\binom{2nm}{2}} = \frac{2n.m!}{2!(m-2)!} \times \frac{2!(2nm-2)!}{(2nm)!} = \frac{1-\frac{1}{m}}{2n-\frac{1}{m}}$$

For $m = 1$ this proportion is 0; for $m = 2$ it amounts to 0.037 for rye and to 0.012 for wheat; for $m = 3$ it amounts to 0.049 for rye and to 0.016 for wheat. For $m \to \infty$ the proportion is,

$\frac{1}{2n}$ *.e.* 0.071 for rye and 0.024 for wheat.

For the case of an even distribution across all chromosomes of the polygenic loci affecting the considered trait it is concluded that the proportion of pairs of linked loci tends to be low. (In an autotetraploid crop the chromosome number amounts to $2n = 4x$. The reader might like to consider what this implies for the above expressions.)

APPLICATIONS OF PLANT GENETIC ENGINEERING

"PHARMING" AND "PLANTIBODIES"

An increasingly viable option is the production of highly valuable enzymes by plants and animals. In addition to production of human proteins in these organisms, other valuable proteins that are currently produced by microorganisms could very well be produced by higher organisms instead. Animal and plant "bioreactors" in some respects may be superior to recombinant bacterial systems, because eukaryotes glycosylate proteins. Whereas the glycosylation pattern may be species-specific, appropriate glycosylation is often required for protein function. Production through these organismal systems may also be cheaper than cell fermentation techniques. Two examples of production of human proteins in plants include the production of human serum albumin in transgenic tobacco and potato, and production of human insulin by tobacco. In both cases, the produced protein appears to be fully effective in humans.

Unfortunately, however, one cannot raise his/her insulin level by eating transgenic tobacco leaves, as the protein in most cases will be broken down to amino acids before it reaches the blood stream. Therefore, in these cases one cannot escape the practice of protein isolation and purification before transgenic leaves are converted into drugs. Also antibodies are being produced in plants. Initially, the antibody's light and heavy chains were produced in different plants. But a subsequent cross of these two varieties resulted in progeny carrying assembled and functional antibodies.

REVERSIBLE MALE STERILITY IN PLANTS

As has been indicated earlier, heterozygous individuals often are healthier and stronger than homozygous ones. The only way to guarantee heterozygoiscity in plants is to make sure self-pollination cannot occur. For most crop plants it was very tedious or practically impossible to exclude selfing. To exclude self-pollination, it would be good to introduce male sterility in plants: progeny from such plants are then expected to be 100 per cent heterozygous (assuming they were pollinated with pollen from an unrelated variety). To introduce male sterility, a promoter was identified that was turned on exclusively in tapetum cells (a tissue around the pollen sac that is essential for pollen production). This promoter then was linked up to a gene coding for a bacterial ribonuclease (named barnase). This ribonuclease selectively chops up ribonucleic acids.

The promoter/ribonuclease construct was then introduced into plants (canola, tobacco, you name it). Because the promoter allows expression only in tapetum cells, the gene construct disrupts only development of the tapetal tissue and its end product, pollen. Plants transformed with this construct were male-sterile but otherwise normal. Although male-sterile plants are valuable for hybrid seed production, they have limited value when it comes to crop production. Fertility must be restored to crops such as wheat, rice, and tomato, in which the seed or fruit is the harvested product. Fortunately, the ribonuclease is inhibited very much by a simple protein, named barstar. One can thus cross the male-sterile plant with a male-fertile variety in which the gene for barstar has been introduced, and the result is progeny with viable pollen and restored fertility. A closely related approach has been criticized as "terminator technology" as it is seen by its critics as a way for companies to protect and enforce their patents.

ANTISENSE RNA

Antisense RNA refers to nucleotide strands that are produced in a cell and that are complementary to a particular mRNA. Antisense RNA can be produced, by inverting the coding region of a gene with respect to its promoter. The antisense RNA can hybridize with its corresponding mRNA, making it double-stranded.

The double-stranded mRNA no longer can be recognized by the protein-synthesizing machinery (the ribosomes), and thus expression of this mRNA is suppressed. Also, in many systems double-stranded mRNA is very unstable and is broken down quickly. Thus, one can inactivate specific genes while not interfering with others.

Antisense approaches already are used to protect plants from damage by plant viruses. Reversal of a gene from bean yellow mosaic virus (BYMV), and putting it into tobacco under a reasonably strong promoter, has led to a tobacco variety that is quite resistant to BYMV. A similar approach is used to transfer viral resistance to other plants. This finding is of significance, in that currently no effective, environmentally friendly methods exist to control many plant viruses.

Very related to this approach is the RNAi (RNA interference) approach. This is very useful for both agriculture and medicine, and the first examples of practical applications of this RNAi technology are appearing. As with any new technology, the initial pilot projects are sort of pedantic. However, more exciting application possibilities abound. Obviously, RNAi technology provides an excellent approach for reverse genetics in eukaryotes. With this method one can turn off genes.

AGRICULTURAL APPLICATIONS IN DEVELOPING COUNTRIES

Perhaps indicative of the large potential and relative ease of genetic engineering, developing countries (particularly China) are progressing rapidly in development and application of genetically engineered crops. Some have gone into commercial production well ahead of similar crops in the US. In China, tomatoes that have been engineered for improved virus resistance have been on the market since late 1992.

There are two main reasons for the more rapid commercialization of bioengineered crops in the developing world:

- Less tight governmental approval mechanisms,
- Hungrier populations.

While in developed countries the main value of biotechnological applications may be to reduce production costs, in the developing world a main factor is the production of more food. Indeed, genetic engineering applications seem to be pretty successful to cut down on pathogen-induced losses. Genetic modification of papaya plants (expression of the ringspot virus coat protein in the plant) protects very well against the very destructive ringspot virus.

PLANT TRANSFORMATION

Genetic engineering of plants is much easier than that of animals.

There are several reasons for this:

- There is a natural transformation system for plants (the bacterium *Agrobacterium tumefaciens*),
- Plant tissue can redifferentiate (a transformed piece of leaf may be regenerated to a whole plant), and
- Plant transformation and regeneration are relatively easy for a variety of plants.

The soil bacterium Agrobacterium tumefaciens ("tumefaciens" meaning tumor-making) can infect wounded plant tissue, transferring a large plasmid, the Ti plasmid, to the plant cell. Part of the Ti (tumor-inducing) plasmid apparently randomly integrates into the chromosome of the plant.

The integrated part of the plasmid contains genes for the synthesis of:

- Food for the bacterium, and
- *Plant hormones*: Genes from the Ti plasmid that are integrated in the plant chromosome are expressed at high levels in the plant.

Overproduction of the plant hormones leads to continuous growth of the transformed cells, causing plant tumors. Rapid, cancerous growth of the transformed plant tissue obviously is advantageous to the bacterium: more food gets produced. The Ti plasmid has been genetically modified ("disarmed") by deleting the genes involved in the production of bacterial food and of plant hormones, and inserting a gene that can be used as a selectable marker. Selectable marker genes generally are coding for proteins involved in breakdown of antibiotics, such as kanamycin. Any gene of interest can be inserted into the Ti plasmid as well. In principle, one can thus transform any plant tissue, and select transformants by screening for antibiotic resistance.

However, unfortunately, there are some complications:

- It has proven difficult to transform some monocots (grasses, etc.) by *Agrobacterium*, and
- Regeneration of plants from tissue culture or leaf discs is not always possible.

A number of genetically engineered plant varieties have been developed. Traits that have been introduced by transformation include herbicide resistance, increased virus tolerance, or decreased sensitivity to insect or pathogen attack. Traditionally, most of such genetically engineered plants were tobacco, petunia, or similar species with a relatively limited agricultural application. However, during the past decade it now has become possible to transform major staples such as corn and rice and to regenerate them to a fertile plant. Increasingly, the transformation procedures used do not depend on *Agrobacterium tumefaciens*. Instead, DNA can be delivered into the cells by small, μm-sized tungsten or gold bullets coated with the DNA.

The bullets are fired from a device that works similar to a shotgun. The modernized device uses a sudden change in pressure of He gas to propel the particles, but the principle of "shooting" the DNA into the cell remains the same. This DNA-delivery device is nicknamed "gene gun", and has been shown to work for DNA delivery into chloroplasts as well. Over the last several years, use of the "gene gun" has become a very common method to transform plants, and has been shown to be applicable to virtually all species investigated.

Transformation of rice by this method is now routine. This is a very important development as rice is the most important crop in the world in terms of the number of people critically dependent on it for a major part of their diet. Another method to get foreign genes into cereals is by electroporation: a jolt of electricity is used to puncture self-repairing holes in protoplasts (*i.e.*, the cell without the cell wall), and DNA can get in through these holes.

However, it is often very difficult to regenerate fertile plants from protoplasts of cereals. Nonetheless, significant advances in overcoming these practical difficulties have been made over the years. Now even transgenic trees have been created: the gene for a coat protein of the plum pox virus has been introduced into apricot. The plum pox virus leads to the feared Sharka disease, for which there is no cure. The resulting transgenic tree shows a markedly decreased sensitivity to this virus. The reason why continuous exposure of the tree to the viral coat protein leads to tolerance against viral infection is not yet understood, however.

Thus, now there are a number of different techniques to introduce foreign genes into plants. Essentially all major crop plants can be (and have been or are being) genetically engineered, the procedures are now routine and the frequency of success is very high. Even though genetically engineered crops are more costly than the usual ones, they have been rather readily accepted by US farmers provided that tangible benefits can be demonstrated. However, it is questionable whether the farmer in poorer countries can come up with the funds to "try out" and use the new crops.

Another issue in this respect is how genetically engineered crops are perceived by the consumer. Even though in the US there is little resistance to such crops as long as the products can be shown to be safe and advantageous, in other countries genetically modified foods are received poorly by the consumer. It is unlikely that there is a rationally sound basis for this rather hostile reaction of the consumer, as most of the crops are the result of human manipulation and may have been treated with harmful herbicides and pesticides.

Time and education will need to be invested to provide consumers and consumer advocates with a balanced opinion on the acceptability of the origin of their foods. One area of particular concern for some people is the lack of labeling of genetically engineered foods, and legislation may be introduced to address this issue.

On the other hand, as so many plants (soybean, corn, etc.) are genetically modified and the nature of the genetic modification is not necessarily easy to explain, it may be simpler to label those foods that are guaranteed free of "genetically modified organisms" or their products. However, keep in mind that essentially all agricultural products have been genetically modified by traditional breeding, so it may be difficult to define what is actually free of genetically modified organisms.

GENE TRANSFER

Once useful genes have been located on a framework map, it is then possible to use the molecular markers to facilitate the transfer of the useful genes between species and strains in an efficient manner. As stated earlier, recombination is not a frequent event along a chromosome and, furthermore, two recombination events rarely occur within a distance of 15 to 20 cM of each other.

It therefore follows that if a useful target gene (a major gene or a QTL) is known to be located within a region of chromosome flanked by two markers that are no more than 15 cM apart, then these markers can be used to follow and control the progression of the target gene through successive stages of a breeding programme. Clearly, the alleles at the marker loci have to differ in the donor and recipient cultivars, but any chromosome that contains A1 and B1 at the marker loci will also contain T1, the target allele to be transferred.

Should single recombination occur close to T, then the chromosome will contain A1 or B1 but not both. Providing the markers are less than 15 cM apart double recombination resulting in A1T2B1, and hence loss of T1, can be safely ignored.

If the location of T is not accurately known, then three or more marker loci may be needed to be sure of safely bracketing the region containing T without fear of double recombination. There are conflicting requirements in marker-assisted gene transfer: the need to keep the bracketed region large enough to be sure of holding the gene to be transferred whilst not having it so large that too many other linked but undesirable alleles are transferred with it. Let us now consider the basic procedure of gene transfer. Traditionally, breeders have introduced a useful allele into their cultivars by backcrossing the original F1 to the commercial cultivar whilst selecting at each generation for the desired allele.

As was shown earlier, this can result in a very large region of chromosome around the target gene surviving even after ten generations of backcrossing – a phenomenon known as linkage drag. With markers, however, it is possible to reduce linkage drag considerably whilst simultaneously reducing the number of backcrossing generations to two or three instead of the normal six to eight. Moreover, the amount of plant material raised at each generation can be reduced because much of the molecular genotyping can be achieved at the juvenile stage. The aim is to hold the target allele, T1, heterozygous throughout the backcrossing process, by selecting for the flanking markers, whilst simultaneously selecting for the genotype of the recurrent, commercial cultivar at all other loci. The latter can readily be achieved by having as few as four or five well-spaced markers on all chromosomes and selecting for recurrent parent alleles at each generation.

Depending on the chromosome number, some individuals in the first backcross generation will be homozygous for several whole chromosomes whilst still being heterozygous for the target gene. Those which have most of the recurrent parent chromosomes are selected and backcrossed again, and if necessary the process is repeated for a third generation. For each generation, fewer markers need to be screened as those found to be homozygous in the previous backcross no longer have to be checked.

Unless the position of the target gene is very accurately defined, it is prudent to have several markers covering its likely position, again with the proviso that they need to be separated by no more than 15 cM. At the final stage of the introgression process, a backcross individual which is homozygous for most of the recurrent parent alleles is selfed or, if this is not possible, intercrossed with a similar genotype.

From among the progeny, individuals are chosen which have different combinations of markers bracketing the target region as shown in figure. These can be selfed again and homozygotes for the different sequences identified and multiplied for reassessment for the target trait. Some will fail to show the target trait because of recombination, but of those that do, the lines with the shortest sequence of donor markers are selected for further trials. The approach can obviously be adapted to the simultaneous introgression of several genes: no new principles are involved but the screening process becomes more elaborate.

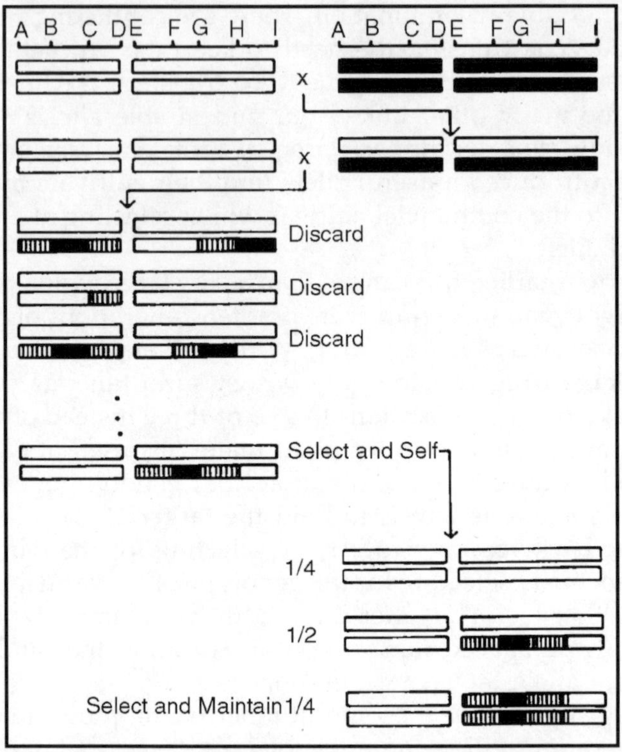

Fig. The use of Molecular Markers to Facilitate Gene Introgression.

It is always possible at a later stage to reduce the length of the introgressed region around T by looking at other, more closely linked, markers. Similarly, it is wise to check the origins of the other chromosomes to avoid having introgressed unwanted regions through missing double recombinants in the central regions of individual chromosomes or single recombinants towards their ends.

If such errors are found they can easily be corrected by another round of backcrossing to the recurrent parent. When attempting to reduce the length of donor chromosome surrounding the target gene, it is important to realise that this should be done in two successive generations. If the sites of the required recombination events are less than 15 cM apart, then the simultaneous double event will not occur. Even if they are far enough apart for the double event to occur, the probability of such an event may be too low to be practicable. If two recombination events each have a probability of 0.05 (*i.e.* 5%), their combined probability is at best 0.0025. Therefore, it would be necessary to genotype approximately 1200 individuals to be 95% sure of having at least one double recombinant.

If it was tackled in two stages then it would require only the genotyping of 86 (= 28 in the first round plus 58 in the second; less being needed in the first round because the recombination could occur on either side of T), a very considerable saving in effort of 93%, or 1114 plants! Marker-aided selection has been successfully tried for quantitative traits in several crops while the efficiency of the procedure has been examined by Lande and Thompson (1990).

MAP-BASED GENE CLONING

Providing a target gene can be located to within a pair of marker loci, it should be possible to use standard cloning techniques to walk along the chromosome between the markers and to find the target gene. The feasibility of this depends on the distance between the markers and the ease with which the target gene can be recognized from among the clones. Depending on the species, a genetic length of 1 cM can, on average, consist of as few as 280 kbp of DNA in rice, or 2220 kbp in maize.

It is not always easy to locate a target gene with sufficient accuracy to be able to say that it is between two markers as close as 1 cM, and hence the actual lengths of DNA between them can often be much greater. With QTLs one can seldom achieve anything approaching this accuracy by existing methods of conventional mapping. Furthermore, these are average distances per centimorgan; some chromosomal regions could be much longer or, indeed, less. Increases in mapping reliability can be obtained by using what are called near isogenic lines (NILs), which are lines obtained by selfing or backcrossing and are known to differ from some standard genotype by just a short defined section of chromosome.

Providing this section is delineated by accurately mapped flanking markers, and the NILs which do and do not contain this region can clearly be shown to differ for the target gene, then the target gene has to be in that region. The genetic and physical proximity of the markers can be established by conventional mapping and by gene cloning. Map-based gene cloning involves starting from one of the markers and walking to the next by means of a series of partially overlapping cosmid clones. Cosmid clones are necessary because they can be up to 50 kb in length and hence require fewer cloning steps in order to cover the DNA between the markers (280 to 2200 kb per cM in the examples cited above).

Identifying the target gene is more difficult. Clearly, if the gene is between the two flanking markers it has to be on one of the cloned regions and could possibly be on two overlapping clones. It could be identified by transforming plants with each clone separately and identifying which transformant expresses the effect. If the gene concerned exhibits classical dominance it would be necessary to transform the homozygous recessive with the dominant allele or the transformant would not be recognized. Alternatively, the gene has to be introduced into a species that does not normally express the gene. Once the effective cloned fragment is identified, the actual location of the gene within the fragment could be identified by successively transforming sub-fractions of the clone or, following sequencing, by looking for potential open reading frames. In those cases where the target gene produces a known product, cDNA clones derived from tissue likely to be rich in the appropriate mRNA of the target gene can be used to identify likely sequences in the previous overlapping clones.

Once the gene has been identified by one of these methods, its structure and activity can be studied in detail. Disease resistance genes are obvious candidates for such studies as their action is very specific and their location on the genome of many species is well documented. However, unless suitable candidate loci.

GENE MAPPING

An important, though by no means an essential, step in genetic analysis is to produce genetic maps of the marker loci. Such maps are often referred to as 'framework maps' because they provide a framework within which important genes can be located, as well as providing a means of comparing chromosome organization in other closely or distantly related species. There are two stages to mapping. Firstly, to arrange the markers in a linear sequence separated by an appropriate map distance, *i.e.* to construct a linkage map. The second is to relate the linkage maps to particular recognizable chromosomes. The latter is often the most difficult and generally not essential for breeding or conservation work. We will, therefore, concentrate on the former. Chromosomes contain a single linear molecule of DNA and hence the markers on that chromosome occur at particular positions along that molecule. Typically, each chromosome contains 107 to 108 base pairs (bp), *i.e.* 104 to 105 kilo-base pairs (kbp) of DNA, while a typical structural gene, coding for a polypeptide chain, would be between 1 and 2 kbp long. Assuming only 10% of the genome is coding DNA, then a chromosome probably contains something of the order of 1000 to 10 000 genes. Fortunately, the fact that the chromosome is a linear molecule means that the genes need to be mapped in only one dimension.

Currently, the only useful method of gene mapping, at least as far as breeders and conservationists are concerned, relies on recombination between homologous chromosomes resulting from genetic exchange, which can be seen as chiasmata at diplotene through first metaphase of meiosis. A single chiasma on a particular chromosome results in half the gametes from such a meiosis being recombinant for that chromosome: half the gametes will contain a copy of that chromosome that is part maternal and part paternal in origin, while the remaining gametes will be entirely parental. Very few plant species have a sufficiently low number of clearly recognizable chromosomes that the number of chiasmata on a particular chromosome can be compared in different nuclei.

However, in many species the total number of chiasmata in each diplotene nucleus can be counted, at least in pollen meiosis, and the average number of chiasmata per nucleus calculated. The number and position of chiasmata on a particular pair of homologous chromosomes will vary from nucleus to nucleus; it is as though there are a large number of potential sites of exchange, but that in any given meiosis only a few sites are actually involved. The longer the chromosome the more potential sites there might be. There will invariably be one chiasma and there may be two, three, four or more; a typical set of results is shown in Table A chromosome that has one chiasma on average is said to be 50 centimorgans (cM) long – a map length unit named after the American geneticist Thomas Hunt Morgan.

This relates to the fact mentioned above that one chiasma results in 50% recombinant chromosomes. By extension, a chromosome with an average of 2.5 chiasmata is said to be 125 cM (*i.e.* 2.5 350 cM) long. It appears that, as a rough rule of thumb, each chromosome has on average two chiasmata and so is 100 cM long. It follows, therefore, that providing the haploid chromosome number is known (*n*), the total map length will be approximately 2 3 *n* 3 50 cM. The average number of chiasmata per nucleus has been calculated in many species and this rule generally holds.

Table. Chiasma Fraquency and Mapping.Chiasma Frequency in Chromosomes of Secale Cereale.

Number of chiasmata per chromosome	Frequency	Percentage unrecombined Chromosomes
1	0.267	50.0
2	0.716	25.0
≥3	0.017	12.5
Mean	–	22.0

This rough rule is important because it gives the geneticist an idea of the total genetic map length that should be expected as more genes or markers are mapped. It provides a guide to the extent to which the currently mapped markers cover the full genetic map. Because the map is based in chiasma units, it is not the same as a physical map. The distribution of chiasmata is not uniform over a given chromosome in a species and will vary with the chromosome and the species. Although a knowledge of the number of chromosomes and, ideally, the mean chiasma frequency per nucleus for the species, indicates the total map length, the actual map has to be constructed from examining the frequencies of progeny in crosses; *i.e.* maps are constructed from information based on the consequences of chiasmata, namely recombination of genetic markers. Recombination of genetic markers is identified by the presence of gametes that contain recombined markers, and such gametes can be recognized from the phenotypes of progeny of a cross. The frequency of such gametes is an estimate of the recombination frequency between the two markers concerned.

The closer two markers are to each other on the chromosome, the less likely a chiasma will occur between them and the less likely they are to recombine. Providing the two markers are sufficiently close that either none or one chiasma occurs between them but never more, then the recombination frequency as a percentage is equal to the map distance in centimorgans as described above. The problem arises when the markers are sufficiently far apart for two or more chiasmata to occur between them in some meioses. It can be shown that not only does a single chiasma between a pair of markers result in 50% recombination between them, but so also do two, three or more chiasmata, on average.

It is necessary to say 'on average', because there are a variety of possible consequences with two, three or more chiasmata in any given meiosis but, in practice, when one is looking at the progeny of a cross or self, each progeny will be the result of gametes from different meioses. This fact implies that although map distance increases linearly with the number of chiasmata, the same is not true with the frequency of recombination, which can never be more than 50% for a pair of markers even though they might be 100 cM apart at opposite ends of a chromosome.

These relationships are illustrated in Figure. In order to overcome this problem, mapping functions have been devised to correct for multiple chiasmata. The most common of these are the Haldane (1919) and Kosambi (1943) mapping functions. Haldane's assumes that the probability of none, one, two or more chiasmata in a given interval follows a Poisson distribution, *i.e.* that the chiasmata are independent, random events. Kosambi's method allows for a certain degree of nonindependence in chiasma occurrences. It has long been known by cytogeneticists that chiasma interference occurs over quite large regions of a chromosome. Interference means that if a chiasma occurs at a particular point on the chromosome, the next one will never occur closer than some fixed distance, the interference distance, from it. Beyond this distance, there is a short distance in which the interference disappears and the next chiasma can then occur randomly outside this range. Observational data in various species of plants are confirming this and suggest that the interference distance is generally between 15 and 20 cM, *i.e.* 15 to 20% of a typical chromosome on either side of the chiasma. This has important and useful consequences for genetic and breeding work as will be shown later.

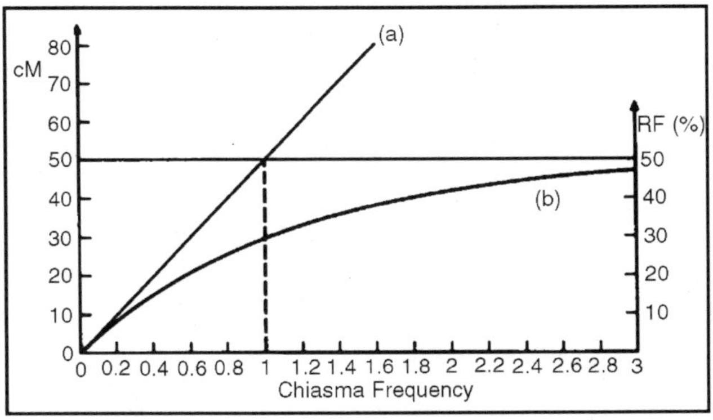

Fig. The Relationship of Chiasma Frequency With (a) Genetic Map Length and(b)Recombination Frequency.

Recombination frequencies between pairs of markers can be scored in a wide variety of different crosses. The simplest to use are generations derived from an F1 because only two alleles are segregating, and their distribution in the chromosomes of the parents of the F1 can be determined. The generations that can be used are F2, backcrosses (Bc), recombinant inbred lines (RILs) or doubled haploid (DH) lines, with F2s being the most informative.

General formulae for calculating recombination frequencies are given by Allard. In outbreeding species where F1s are not available, a given individual can be considered as an F1 and its selfed progeny as being an F2, even though the inbred parents do not exist. If it cannot be selfed but controlled crossing to another individual is possible, then again those genes segregating in the cross can be mapped although the situation is more complex.

There could be from two to four alleles segregating at each locus and, with respect to any one locus, the cross could represent an F2, Bc1 or Bc2. Moreover, the distribution of alleles in the two parents cannot be known and has to be inferred from the progeny. This complexity is compounded by the large number of marker loci that may be segregating in any given cross. Fortunately, a range of software packages are available to estimate these recombination frequencies, identify linkage groups, assign the markers to the most likely order and space them in map units (cM) on these linkage groups; examples are MAPMAKER and JOINMAP. Ideally the number of linkage groups should be equal to the chromosome number in the gametes but, unless the markers available provide a good coverage of the genome, it is likely that those markers on a given chromosome may appear as two or more separate linkage groups simply because the subsets are not sufficiently close for them to be recognized as being together on one chromosome.

A typical marker framework map is shown in Figure for the chromosomes of *Brassica oleracea*. Clearly these maps provide a very detailed coverage of the chromosome. Many of the markers are very close, *i.e.* less than 10 cM, and over this distance it matters little which of the two mapping functions is used.

With the more widely spaced markers, *i.e.* recombination frequency >15%, Haldane's function will exaggerate their distance apart because it will allow for more double crossovers than actually occur. As more markers are mapped, the total length should converge on the value predicted by the chiasma frequency, *i.e.* approximately 2 3 n 3 50 cM.

It is important to remember that genetic map distances have standard errors which are dependent on the size and type of population used to construct the map, the mapping population. The map obtained is that which best fits, given the data and the assumptions underlying it, but there may well be other maps which also fit the data well, though are slightly less likely. Any particular map always develops a greater aura of respectability once it is published, frequently without stating its reliability, and subsequent yet different maps are treated with suspicion. A recombination frequency of true value p has a standard error of ?[p(1 2 p)/N], where N is the size of the gamete population. For example, if p is 0.1 (*i.e.* RF = 10%) and$N$ = 100, the standard error is 0.03 or 3%. In other words the 95% confidence interval of the estimate lies approximately between 4% and 16%.

There are other reasons, apart from statistical sampling, why maps could be wrong. The most obvious is the accuracy with which the data are scored and recorded. Autoradiographs and banding patterns on gels can easily be misread and research workers are loath to discard data even though its interpretation is ambiguous. All data should be checked independently by two or more people and any ambiguities removed. Wrong scores will bias the data and often exaggerate map lengths as they suggest double recombination; wherever double recombination appears to have occurred in two short, adjacent intervals, the data should be checked. Changes in methylation rather than scoring errors could cause a restriction site to appear or disappear so giving the false impression of double crossing- over, or it could simply be an error.

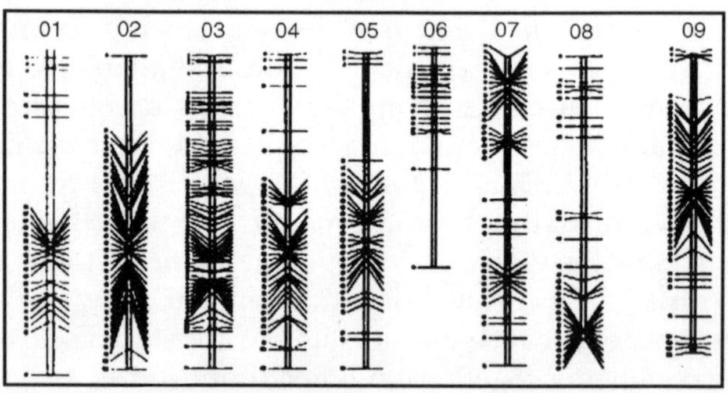

Fig. Typical Molecular Marker Framework Map.The Nine Groups of Brassica Oleracea Based on RFLP, Isozyme and Morphological Markers

Where maps appear to have major inconsistencies in different crosses, different chromosomal structural arrangements may be responsible, such as translocations or inversions. There is considerable evidence that chiasma frequencies, and hence recombination frequencies, may be quite different in male and female meioses even on the same plant although in other studies this may not be so.

Thus although the order of markers on a linkage group may remain the same, the relative distances between them may be quite different if the map derives from male versus female meioses. For example, in *Brassica* it appears that the genetic map obtained from a backcross where the F1 is the female parent is 60% longer than when the F1 is the male parent. It is also clear that environmental factors during meiosis, particularly temperature, can affect the number and distribution of chiasmata, and this could be very critical if the crosses are set up at different times of the year.

Two or more different populations will almost certainly be segregating for different combinations of polymorphic markers, although some at least should be in common. These common ones can be used to overlay the two or more maps and a consensus map produced from an amalgamation of these. Again software is available to produce these consensus maps. Despite the various error-causing factors discussed above a considerable degree of consensus can be produced from such populations and, considerable similarity of chromosome sequence is conserved between even distantly related species, allowing the potential for cross-species genetic transfer.

It was stated above that the total map length should be dictated by the chiasma frequency, in so far as this can be accurately measured. When the first few genes were initially put onto genetic maps in the first 40 years of this century, they only covered a small part of the genome – except in wellstudied species such as *Drosophila* and maize. As more and more genes were placed on the map, so the total map lengths increased towards the asymptote dictated by the chiasma frequency. However, during the first decade following the use of molecular markers, the map lengths of some species appeared to exceed that expected, and it was even argued by some that the chiasma theory of recombination might be wrong.

No one would wish to argue that chiasma frequency data are free from error but there would have had to have been quite excessive underscoring to result in the disparity which was occasionally found. It would appear, however, that the excessive lengths were due in part to errors in scoring and to the use of small mapping populations, because subsequent map sizes have shown a progressive decrease towards the predicted asymptote. Scoring errors bias estimates of RF generally upwards, while RFs have to be large to be detectable in small populations.

Our own experience with mapping in cereals and brassicas has also shown a progressive decrease in estimated map length as more markers can be found to fill some of the large gaps in the map and as errors are removed from the mapping data.

GENETIC ALGORITHMS

Genetic programming is a branch of genetic algorithms. The main difference between genetic programming and genetic algorithms is the representation of the solution.

Genetic programming creates computer programmes in the lisp or scheme computer languages as the solution. Genetic algorithms create a string of numbers that represent the solution.

Genetic programming, uses four steps to solve problems:

1. Generate an initial population of random compositions of the functions and terminals of the problem (computer programmes).
2. Execute each programme in the population and assign it a fitness value according to how well it solves the problem.
3. Create a new population of computer programmes.
   - Copy the best existing programmes
   - Create new computer programmes by mutation.
   - Create new computer programmes by crossover(sexual reproduction).
4. The best computer programme that appeared in any generation, the best-so-far solution, is designated as the result of genetic programming.

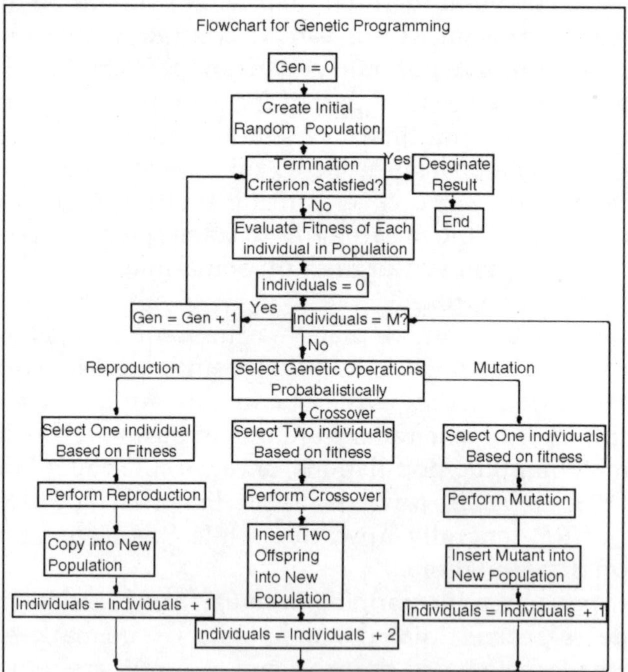

Fig. Genetic Programming Flowchart.

FITNESS FUNCTION

The most difficult and most important concept of genetic programming is the fitness function. The fitness function determines how well a programme is able to solve the problem. It varies greatly from one type of programme to the next. For example, if one were to create a genetic programme to set the time of a clock, the fitness function would simply be the amount of time that the clock is wrong. Unfortunately, few problems have such an easy fitness function; most cases require a slight modification of the problem in order to find the fitness.

GUN FIRING PROGRAMME

A more complicated example consists of training a genetic programme to fire a gun to hit a moving target. The fitness function is the distance that the bullet is off from the target. The programme has to learn to take into account a number of variables, such as wind velocity, type of gun used, distance to the target, height of the target, velocity and acceleration of the target. This problem represents the type of problem for which genetic programmes are best. It is a simple fitness function with a large number of variables.

WATER SPRINKLER SYSTEM

Consider a programme to control the flow of water through a system of water sprinklers. The fitness function is the correct amount of water evenly distributed over the surface. Unfortunately, there is no one variable encompassing this measurement. Thus, the problem must be modified to find a numerical fitness. One possible solution is placing water-collecting measuring devices at certain intervals on the surface.

The fitness could then be the standard deviation in water level from all the measuring devices. Another possible fitness measure could be the difference between the lowest measured water level and the ideal amount of water; however, this number would not account in any way the water marks at other measuring devices, which may not be at the ideal mark.

MAZE SOLVING PROGRAMME

If one were to create a programme to find the solution to a maze, first, the programme would have to be trained with several known mazes. The ideal solution from the start to finish of the maze would be described by a path of dots. The fitness in this case would be the number of dots the programme is able to find. In order to prevent the programme from wandering around the maze too long, a time limit is implemented along with the fitness function.

FUNCTIONS AND TERMINALS

The terminal and function sets are also important components of genetic programming. The terminal and function sets are the alphabet of the programmes to be made. The terminal set consists of the variables and constants of the programmes. In the maze example, the terminal set would contain three commands: forward, right and left.

The function set consists of the functions of the programme. In the maze example the function set would contain: If "dot" then do x else do y. In the gun firing programme is the terminal set would be composed of the different variables of the problem.

Some of these variables could be the velocities and accelerations of the gun, the bullet and target. The functions are several mathematical functions, such as addition, subtraction, division, multiplication and other more complex functions.

CROSSOVER OPERATION

Two primary operations exist for modifying structures in genetic programming. The most important one is the crossover operation. In the crossover operation, two solutions are sexually combined to form two new solutions or offspring.

The parents are chosen from the population by a function of the fitness of the solutions. Three methods exist for selecting the solutions for the crossover operation. The first method uses probability based on the fitness of the solution. If f $(s_i(t))$ is the fitness of the solution Si and,

$$\sum_{j=1}^{M} f\left(s_j(t)\right)$$

is the total sum of all the members of the population, then the probability that the solution Si will be copied to the next generation:

$$\frac{f\left(s_i(t)\right)}{\sum_{j=1}^{M} f\left(s_j(t)\right)}$$

Another method for selecting the solution to be copied is tournament selection. Typically the genetic programme chooses two solutions random. The solution with the higher fitness will win.

This method simulates biological mating patterns in which, two members of the same sex compete to mate with a third one of a different sex. Finally, the third method is done by rank. In rank selection, selection is based on the rank, (not the numerical value) of the fitness values of the solutions of the population. The creation of the offsprings from the crossover operation is accomplish by deleting the crossover fragment of the first parent and then inserting the crossover fragment of the second parent.

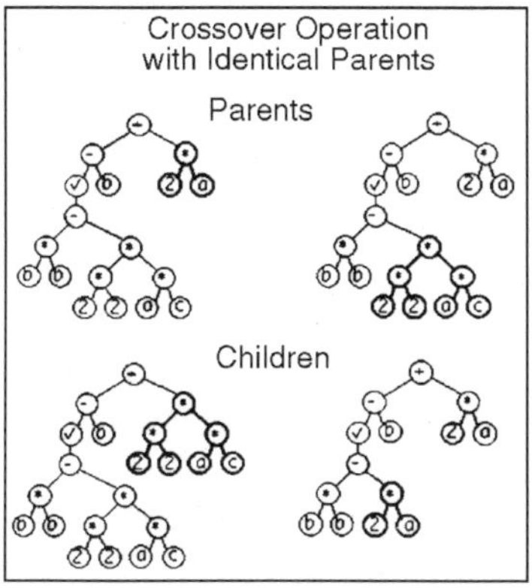

Fig. One of the Main Advantages of Genetic Programming Over Genetic Algorithms.

In genetic programming identical parents can yield different offspring, while in genetic algorithms identical parents would yield identicaloffspring. The bold selections indicate the subtrees to be swapped. An important improvement that genetic programming displays over genetic algorithms is its ability to create two new solutions from the same solution. In the Figure the same parent is used twice to create two new children.

MUTATION

Mutation is another important feature of genetic programming. Two types of mutations are possible. In the first kind a function can only replace a function or a terminal can only replace a terminal. In the second kind an entire subtree can replace another subtree. Figure explains the concept of mutation:

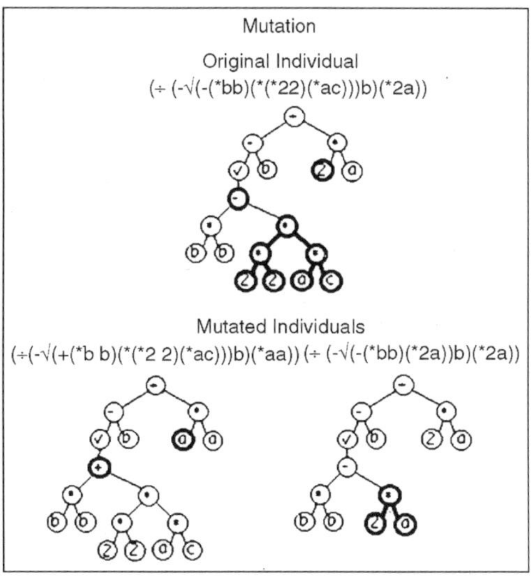

Fig. Two Different Types of Mutations.

The top parse tree is the original agent. The mutation of a single terminal for another single terminal (a). It also illustrates a mutation of a single function (–) for another single function (+). The parse tree on the bottom right illustrates a the replacement of a subtree by another subtree. Genetic programming is much more powerful than genetic algorithms. The output of the s is a quantity, while the output of the genetic programming is a another computer programme.

In essence, this is the beginning of computer programmes that programme themselves. Genetic programming works best for several types of problems. The first type is where there is no ideal solution, (for example, a programme that drives a car). There is no one solution to driving a car. Some solutions drive safely at the expense of time, while others drive fast at a high safety risk.

Therefore, driving a car consists of making compromises of speed versus safety, as well as many other variables. In this case genetic programming will find a solution that attempts to compromise and be the most efficient solution from a large list of variables. Furthermore, genetic programming is useful in finding solutions where the variables are constantly changing. In the previous car example, the programme will find one solution for a smooth concrete highway, while it will find a totally different solution for a rough unpaved road.

GENETIC ALGORITHMS - NATURE'S WAY

In the vast eons of time the processes of evolution have created a common basis for all life on Earth - that of DNA or its simpler variant RNA. Both systems share a method of coding with that greatest of human inventions, the computer, yet the mode of operation of the two processes could hardly be more different.

Approaches to Achievement

In our use of computers we first decide an overall goal, we then break down the problem into manageable chunks and write step by step procedures (algorithms) by which to accomplish them. This mirrors our normal approach to life, in which we decide what we wish to do and plan out a way of achieving our ambition, taking into account the resources and time needed. If the plan doesn't work we modify it and try again - a serial, one at a time approach to achievement.

Nature has a different way, brute force and ignorance. If there are a hundred possible options available, well try them all - the first to get there wins. All options are tried simultaneously - a parallel approach to achievement. But where is there ? It just doesn't matter ! At first sight, to human eyes, this is a strange concept - how can you get anywhere useful without knowing where it is ?

TEACHING VERSUS EXPLORING

"Ignorance is bliss" goes the saying and we humans certainly start from a state of ignorance. Each of us can only plan for what we already know, the constraints of our limited knowledge restrict what we can imagine - we often cannot see beyond our assumptions. To overcome these limitations we must rely on being taught new ideas by someone whom we trust, or by stumbling onto new discoveries by accident - in both ways we expand our mental horizons and thus see new possibilities that were previously unthinkable. By adopting nature's method and trying every possible option we can arrive at every destination simultaneously. Thus we can enjoy every possible experience.

Wow, what a lifestyle ! Is there a snag ? Of course. We consume a hundred times as many resources for every hundred possible options - and the options increase geometrically (like sand grains doubling for every square on a chessboard), the universe just isn't big enough. Note however, that the conformist social equivalent of everybody doing exactly the same thing (all making the same mistake...), uses just as many resources, yet achieves nothing better - despite the considerableduplication waste involved, our planet isn't big enough for this either !

EVALUATING OPTIONS

But nature has the solution already, tree pruning. Imagine we have two options, each of those has two more and so on, drawing those out we get a picture like a branching tree.

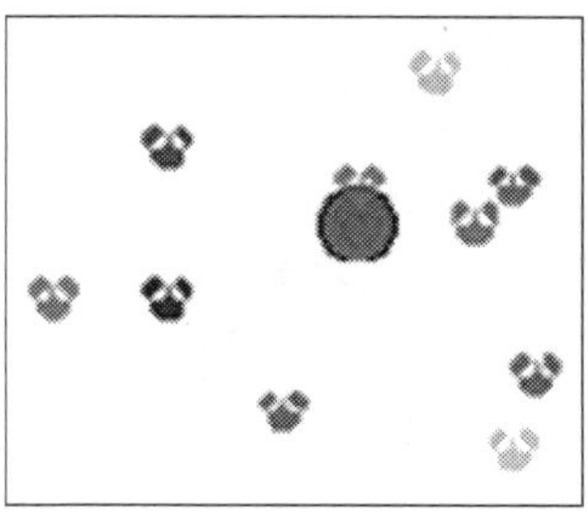

Suppose we impose a test on each option, if the test fails then that option (branch) is deleted (in this illustration we eliminate any mouse genes that fail to find the central food). Once we do this we have pruned the tree of all failed branches and those left passed all our tests (this is similar to how Chess computer programmes work). But which test does nature use ? A very simple one - survival itself. Any organisms unable to survive and reproduce will die out, their genes pruned from the available options. Can computers operate as nature does ? Yes, this is the general field known as Evolutionary Computation, of which the Genetic Algorithm is the basic technique and can be further expanded to remove the need for an underlying programme.

THE GENETIC ALGORITHM

How does it work? Well, we create a population of individuals, each represented by a chromosome (a collection of genes or characteristics) appropriate to the problem we are trying to investigate. Genes are usually implemented by a series of computer bits, enough to cover all available alternative values. The value of each gene is initially assigned randomly, within the available parameters.

The population is then evaluated to determine how well each individual does the required task. The best members of the population (typically the top 10per cent) then become the parents for the next generation. The genes of the offspring are created by selecting two parents at random and combining part of the chromosome of each, so parent gene combinations ABC mated with DEF may become two offspring ABF and DEC. This mimics the crossover or recombination of sexual reproduction in nature and is repeated for further pairs until a full second population is created. Random changes to individual bits are then made mimicking the natural role of mutation, at some desired rate (say 1 in 100 bits changed).

The resultant population is then evaluated as before, and the process is repeated as many times as desired until the required performance level has been achieved (or no further improvement seems possible). This technique can rapidly covers the space of all possible options and converge on a solution that is beyond the ability of all but the best human programmers, in areas where no conventional solution techniques exist.

COEVOLUTIONARY PROGRAMMING

In the Genetic Programming extension, the genes are the valid statements within a computer language and the chromosome in this case is itself the programme. Additional operations are allowed to join or split programme segments, allowing variable length programmes to evolve. By GA techniques (excluding mutation) we can thus allow a programme to create itself to solve a particular problem, discarding all the failed programmes. This has proved a very powerful way of generating solutions to very difficult problems, but with the drawback that the resultant code is all but incomprehensible to humans !

There is a difficulty here. If our test is too specific, then the programme might be perfect for that one task, but fail completely for very similar tasks. How can we overcome this ? Again by copying nature. Suppose we wish to create a programme to sort a series of numbers. If we have a fixed test sequence of numbers we encounter the problem mentioned, so let us have many sets of numbers, as many as we have candidate programmes.

Programmes can then encounter different tests at each stage and only those capable of a general solution will survive. But we can do better still, let us evolve the tests also by the same process, the test survivors will the ones that defeat a programme. Using this coevolution strategy, the tests get harder as the programmes get better and this bootstrapping feedback technique gives a considerable improvement in speed and efficiency. If rather than programmes and tests we have interacting agents we enter the realm of Artificial Life.

Evolutionary Thinking

Can we use these techniques in human thought processes ? Perhaps we already do. Whenever we think of a task do we not automatically generate many possible solutions and then evaluate them (subconsciously in most cases) against our goal, discarding all those that fail our criteria ? We then refine the survivors (combining the best parts of each) and so on until we have an optimum solution. Nature has again beaten us to it...

Socially a similar process is common. Brainstorming is a popular management technique for generating diverse pseudo-random solutions to a problem, equivalent to nature's mix and match juggling of genes perhaps. The group then evaluates (tests) the options and prunes those failing some criteria, the winners then propagate to the next stage and perhaps ultimately into production. Along with top down (analysis) and bottom up (synthesis) strategies - both goal directed, we also operate a parallel evolution strategy, companies (alternative chromosomes) compete to grow, ideas (memes - cultural genes) compete for attention. Evolution is embedded in our behaviour.

ART AND INTERACTIONS

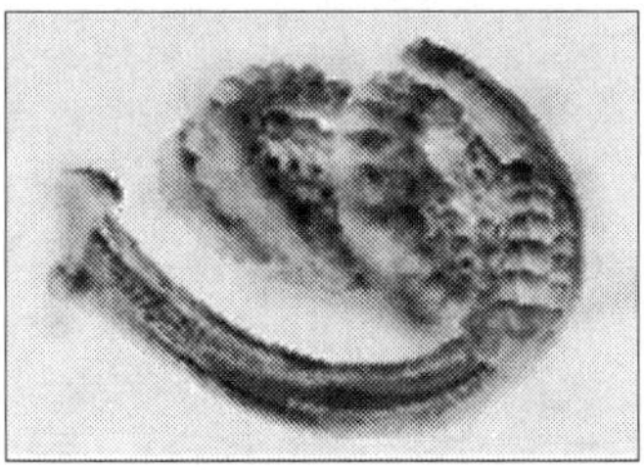

The technique of Genetic Algorithms is widely applicable, even in art. If we code genes to refer to some feature of an image or music we can evolve extraordinary organically based artforms or music. Any system of building blocks can be readily adapted to these procedures and in this way transcend automatically the imagination of its creator. By constraining the system by the desired criteria (tests) we can allow design processes to evolve as a background task - resulting in quite unexpected innovations.

Nature's way. We must however be clear that we are not fully imitating nature here. Our artificial genes (called G-Types) have well defined correspondences with function or form (called P-Types). We also have only one chromosome of very limited length in our genome, whereas natural genotypes contain many chromosomes, each with thousands of genes.In nature genes translate to body form (phenotype or morphology) via a very complex interactive process, the form is itself an emergent phenomena dependent on the interaction of genes with themselves, with the cell chemistry and with the surrounding environment.

We cannot predict this process adequately without understanding the nature of emergence. Genetic engineering, whilst highly useful, is a very dangerous process. The side effects of any changes to a such a complex and highly non-linear process as life are, to say the least, unpredictable.

GENETIC PROGRAMMING OPERATORS

The four general genetic operators are: crossover, reproduction, mutation, and inversion. The following overview describes each in brief.

- *Crossover*: Koza considers crossover, along with reproduction, to be the two foremost genetic operations. It is mainly responsible for the genetic diversity in the population of programmes. Similar to its performance under genetic algorithms, crossover operates on two programmes (a binary operator), and produces two child programmes. Two random nodes are selected from within each programme and then the resultant "sub-trees" are swapped, generating two new programmes. These new programmes become part of the next generation of programmes to be evaluated.

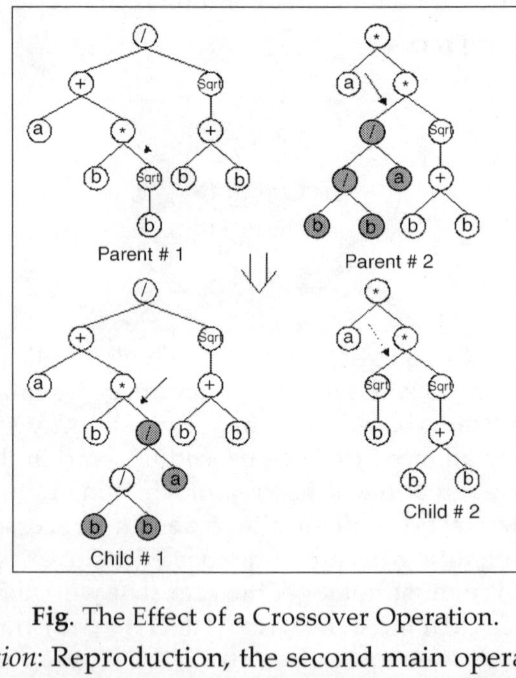

Fig. The Effect of a Crossover Operation.

- *Reproduction*: Reproduction, the second main operation, is performed by simply copying a selected member from the current generation to the next generation. An interesting note about convergence of the population using the two main operators is made by Koza ... in genetic programming, when an individual incestuously mates with itself (or copies of itself), the two resulting offspring will, in general, be different.... As before, the Darwinian reproduction operation creates a tendency towards convergence; however, in genetic programming, the crossover operation exerts a counterbalancing pressure away from convergence. Thus, convergence of the population is unlikely in genetic programming.

- *Mutation*: With genetic algorithms, mutation becomes an important operator which provides diversity to the population. As Koza points out, however, mutation is relatively unimportant in the new environment, because the dynamic sizes and shapes of the individuals in the population already provide diversity, and as stated above, the population should not converge. Thus, mutation can be considered as a variation on the crossover operation.

GENE REGULATION

The processes that enable information to be copied from genes and then used to synthesize proteins must be regulated if an organism is to survive. Different cells within an organism share the same set of chromosomes. In each cell some genes are active while others are not.

For example, in humans only red blood cells manufacture the protein hemoglobin and only pancreas cells make the digestive enzyme known as trypsin, even though both types of cells contain the genes to produce both hemoglobin and trypsin. Each cell produces different proteins according to its needs so that it does not waste energy by producing proteins that will not be used.

A variety of mechanisms regulate gene activity in cells. One method involves turning on or off gene transcription, sometimes by blocking the action of RNA polymerase, an enzyme that initiates transcription. Gene regulation may also involve mechanisms that slow or speed the rate of transcription, using specialized regulatory proteins that bind to DNA. Depending on an organism's particular needs, one regulatory protein may spur transcription for a particular protein, and later, another regulatory protein may slow or halt transcription.

Prokaryotes

The bacterium *Escherichia coli* (commonly referred to as *E. coli*), found in the intestines of humans and other mammals, provides a good example of gene regulation. *E. coli* uses three enzymes to digest lactose, the primary sugar found in milk. The bacterium produces great quantities of lactose-digesting enzymes when lactose is present and saves energy by not synthesizing the enzymes when the sugar is not available. *E. coli* prefers consuming glucose to lactose, so the bacterium produces these enzymes in the presence of lactose only when no glucose is available. A region of DNA known as an operon controls this gene regulation process. In *E. coli*, the operon includes at least five genes: Three genes, called lac genes, code for the enzymes that digest lactose; one gene encodes for a regulatory protein, called a repressor, that can sense the presence or absence of lactose; and one gene, called an operator, activates transcription, the first step in the synthesis of the lactose-digesting enzymes.

In the absence of lactose, the repressor protein binds with the operator gene to block transcription. This prevents the lac genes from being transcribed and halts production of the lactose-digesting enzymes. If lactose is present, the repressor protein binds to the sugar, leaving the operator gene free to trigger transcription of the lac genes. The transcription of the lac genes produces the mRNA that will direct the production of the three lactose-digesting enzymes.

Eukaryotes

Gene regulation in eukaryotes is more complex than in bacteria and other prokaryotes. Even the simplest eukaryotes have far more genes than prokaryotes, and these genes must be turned on or off as conditions dictate. Most multicellular organisms contain different types of cells that serve specialized functions. The cells of an animal's heart, blood, skin, liver, and muscles all contain the same genes. But in order to carry out their specific functions within the body, each cell must produce different proteins and respond to changing environmental stimuli, such as glucose levels in the blood or body temperature. Such specialization is possible only with sophisticated gene regulation. Eukaryotes use a variety of mechanisms to ensure that each cell uses the exact proteins it needs at any given moment. In one method, eukaryotic cells use DNA sequences called enhancers to stimulate the transcription of genes located far away from the point on the chromosome where transcription occurs. If a specific protein binds to an enhancer site on the DNA, it causes the DNA to fold so that the enhancer site is brought closer to the site where transcription occurs. This action can activate or speed up transcription in the genes surrounding the enhancer site, thereby affecting the type and quantity of proteins the cell will produce. Enhancers often exert their effects on large groups of related genes, such as the genes that produce the set of proteins that form a muscle cell.

Gene regulation can also take place after transcription has occurred by interfering with the steps that modify mRNA before it leaves the nucleus to take part in translation. This process typically involves removing exons (segments that code for specific proteins) and introns. These sections of the mRNA can be modified in more than one way, enabling a cell to synthesize different proteins depending on its needs.

GENES IN DEVELOPMENT

Gene regulation helps individual cells within an organism function in a specialized way. Other regulatory mechanisms coordinate the genes that determine how cells develop. All of the specialized cells in an organism, including those of the skin, muscle, bone, liver, and brain, derive from identical copies of a single fertilized egg cell. Each of these cells has the exact same DNA as the original cell, even though they have vastly different appearances and functions. Genes dictate how these cells specialize.

Early in an organism's embryonic development the overall body plan forms. Individual cells commit to a particular layer and region of the embryo, often migrating from one location to another to do so. As the organism grows, cells become part of a particular body organ or tissue, such as skin or muscle. Ultimately, most cells become highly specialized—not only to develop into a neuron rather than a muscle cell, for example, but to become a sensory neuron instead of a motor neuron. This process of specialization is called differentiation. At each stage of the differentiation process, specific genes known as developmental control genes actively turn on and switch off the genes that differentiate cells.

One class of developmental control genes, known as homeotic genes, directs the formation of particular body parts. Activating one set of homeotic genes instructs part of an embryo to develop into a leg, for example, while another set initiates the formation of the head. If a homeotic gene becomes altered or damaged, an organism's body development can be dramatically disrupted. A change in a single gene in some insects, for instance, can cause a leg to grow where an antenna belongs. Homeotic genes work by regulating the activity of other genes. Homeotic genes code for the production of a regulatory protein that can bind to DNA and thus affect the transcription of one or more genes. This enables homeotic genes to initiate or halt the development and specialization of characteristics in an organism.

Nearly identical homeotic genes have been identified in varied organisms, such as insects, worms, mice, birds, and humans, where they serve similar embryonic development functions. Scientists theorize that homeotic genes first appeared in a single ancestor common to all these organisms. Sometime in evolutionary history, these organisms diverged from their common ancestor, but the homeotic genes continued to be passed down through generations virtually unchanged during the evolution of these new organisms.

SCIENTISTS WORK WITH GENES

Scientists have developed a number of biochemical and genetic techniques by which DNA can be separated, rearranged, and transferred from one cell to another. Some of these laboratory methods help scientists study the properties of genes in nature—for example, by comparing DNA from different animals to find out whether those animals are closely related to each other or only distant relatives. Other DNA techniques provide tools for genetic engineering—the alteration of genes in an organism. These tools are used in industry to develop commercial products, such as hardier crops, microbes that can break down oil slicks or decompose garbage, and improved medicines.

RECOMBINANT DNA

The DNA molecules of all life forms, from oak trees to sea horses, have the same structure and the same four bases. Scientists have made use of these similarities in a technology called recombinant DNA. In this laboratory method, one or more genes of an organism are introduced into a second organism. The new genes, sometimes known as foreign DNA, become functional in the second organism and produce a desired protein. In this way, scientists can create changes in the genetic makeup of an organism that would be unlikely to occur through natural processes.

Scientists use recombinant DNA when they want to obtain large amounts of a protein, such as insulin, produced by a gene. Insulin was once in short supply for diabetics, whose bodies lack adequate supplies. Insulin supplies were derived from cows in an expensive and time-consuming process. Today recombinant DNA techniques produce insulin cheaply and in abundance. The first step in creating insulin using recombinant DNA is to isolate the sequence of nucleotides in the DNA of a human cell that forms the insulin gene. Scientists use *restriction enzymes,* specialized proteins that act like molecular scissors, to cut the double-stranded DNA at the point where the insulin gene occurs. The isolated DNA can then be recombined, or spliced, with a vector, a fragment of DNA that is able to transport genes from one organism to another. A vector may be a plasmid, a small, circular segment of DNA found in bacteria. Bacteriophages, viruses that are parasites of bacteria, also act as vectors. Scientists insert the vector containing the insulin gene into a bacterium, such as *E. coli*. Within just a few hours, a single *E. coli* will reproduce hundreds of times to make millions of cells, all containing exact copies of the insulin-producing gene inserted by the scientists. This process of making many cells with identical DNA is known as cloning.

DNA LIBRARIES

A DNA library is a storehouse of genetic information maintained in bacteria instead of books. These bacteria are clones created by recombinant DNA, and the foreign DNA they hold is the library's store of information. DNA libraries are helpful to scientists who require a plentiful supply of particular DNA segments to do their work. These repositories of genetic information are stored in small tubes, which can easily be shipped to other researchers for study. Each library has a unifying theme. For example, a library may contain the entire chromosomal DNA, or genome, of a given organism, or it may consist of genes that are active within certain types of cells, such as heart cells. To create a library of the human genome, DNA from all the human chromosomes would be cut into many pieces. These pieces would be randomly inserted into vectors, such as plasmids, which would then be placed into a population of bacteria. Taken together, the entire population of bacteria would contain all the DNA of the human chromosomes.

POLYMERASE CHAIN REACTION

Polymerase chain reaction (PCR) offers an alternative to vector-based cloning as a means of generating numerous copies of DNA from a small initial sample. Performed in a test tube, PCR mirrors the way in which DNA is replicated within a cell. To perform PCR, scientists isolate the piece of DNA to be amplified (multiplied) in a test tube and heat it to separate the two strands of the molecule. As cooling occurs, short pieces of DNA called primers are added to the test tube.

The primers attach to each strand, marking the segment that will be cloned. Free-floating nucleotides and an enzyme called DNA polymerase are then added to the mixture. DNA polymerase uses the free-floating nucleotides to build a complementary copy of each amplified DNA segment, resulting in two new double-stranded DNA molecules.

Each cycle of heating and cooling doubles the amount of the desired DNA fragment in the test tube. In a matter of hours, scientists can obtain millions of copies of a desired piece of DNA. PCR enables scientists to amplify traces of DNA found at a crime scene or in a fossil animal to produce sufficient quantities to study.

GEL ELECTROPHORESIS

PCR and recombinant DNA techniques create large amounts of DNA segments. To study the structure of these segments, researchers use a process known as gel electrophoresis. This technique can be used to identify genes in humans that have previously been identified in other organisms, such as fruit flies. It can also be used to compare the DNA found from blood or hair samples at a crime scene with the DNA of a suspect in the crime. In gel electrophoresis, restriction enzymes break up the DNA under study into restriction fragments of varying lengths. Solutions containing these fragments are placed within a thick gel.

An electric current is applied to the gel, causing one end of the gel to have a positive charge and the other to have a negative charge. All of the restriction fragments begin to move from the negative end of the gel towards the positive end. The smaller fragments move faster than the larger fragments. When the current shuts off, typically after several hours, the DNA fragments have spread out across the gel, with the smaller ones closer to the positive end. The dispersed fragments display a pattern resembling a bar code. Each bar in this pattern contains DNA fragments of a certain size. Scientists can identify specific restriction fragments by their location on the gel. A complementary sequence of DNA can be used as a probe to find a restriction fragment on the gel that has a particular nucleotide sequence. Scientists may use DNA found in blood at a crime scene as the probe to see if it pairs up with any of the DNA fragments in the gel electrophoresis. If pairing occurs, the DNA from the crime scene is from the same person who provided the DNA sample for the gel electrophoresis.

DNA SEQUENCING

Once an interesting piece of DNA has been isolated or identified, scientists often need to determine if the sequence of nucleotides in the fragment is related to known genes and to determine what kind of protein it might make. Scientists use DNA sequencing to detect genetic mutations linked to diseases such as cystic fibrosis. Scientists have also used this method to alter the sequence of a gene and study the function of the resulting protein. In DNA sequencing, scientists create many copies of a single-stranded DNA fragment that will be used to synthesize a new DNA strand.

An equal number of copies of the fragment are placed into four different test tubes to act as the template for the synthesis of a new strand. The enzyme DNA polymerase and free nucleotides are added to each test tube. Each test tube also receives one type of dideoxy nucleotide—a nucleotide that closely resembles either adenine, guanine, thymine, or cytosine. These nucleotides can attach to the end of the new complementary DNA strand, but they cannot bind to anything else, thus they terminate the synthesis of the new DNA strand.

DNA polymerase uses the free nucleotides to build a complementary DNA strand. If the original DNA fragment contains guanine, DNA polymerase delivers a cytosine dideoxy nucleotide to pair with the guanine base on the original strand. The cytosine links with the growing chain of nucleotides on the complementary DNA strand, but it is unable to bind with any other nucleotide. The newly formed DNA fragment terminates with the cytosine dideoxy nucleotide at the end of the chain. The reactions in each of the four test tubes produce a series of DNA fragments in which the new strands terminate at a known base.

Each test tube produces fragments that differ in length from the other test tubes. The newly formed fragments are sorted in an electrophoresis gel that can detect differences as small as one nucleotide in length. By analyzing these sorted fragments, scientists can determine the complementary base sequence for the original DNA fragment. This sequencing method has become a routine laboratory technique, automated with specialized machines and computers that can prepare DNA samples and read nucleotide sequences far faster and more accurately than people can.

GENE CHIP

The gene chip, also known as a DNA chip or DNA microarray, is a thumbnail-sized chip of glass or silicon that carries DNA instead of electronic circuits. Gene chips can identify the genes that are active within a cell and help identify mutated genes. In one application, scientists take a single strand of DNA that contains a defective gene and use ultraviolet light to attach the strand onto a glass or silicon chip.

A second DNA strand isolated from a patient is attached to fluorescent markers and deposited onto the chip. If the patient's DNA strand bonds with the DNA already bonded to the chip, then the individual's DNA contains the defective gene. When the DNA on the chip pairs with the fluorescent DNA, it develops a fluorescent glow that can be viewed with a microscope and interpreted by a computer. A diagnostic gene chip may soon be manufactured to hold the DNA sequences of all the known disease-causing genes, making diagnosis for genetic disorders fast, reliable, and inexpensive. Gene chips also distinguish between active DNA—DNA that is being transcribed to produce mRNA—and inactive DNA. Researchers use these chips to learn how the transcription of a group of genes is affected when cells are exposed to a drug.

HUMAN GENETICS

Our understanding of human genetics builds on a foundation of information obtained from studying other organisms. Until the 1980s, genetic researchers focused their work on the fundamental genetic processes in simpler organisms, such as bacteria, plants, and fruit flies. Today an expanded array of tools available for the direct study of human genetics attracts scientists from around the world to collaborate to identify and study every human gene. The genetic principles that Mendel first discovered in plants apply to humans as well. As in all other life forms, the DNA found in human cells encodes the proteins that are essential for reproduction, survival, and growth. The unique structure and behaviour of DNA ensures that human traits are passed from generation to generation and accounts for why parents, children, and grandchildren often have similar facial features, hair colour, height, and athletic or artistic abilities. Yet each of us inherits a unique genetic legacy from our parents and more distant ancestors. With the exception of identical twins, no two people have the exact same combination of alleles for the estimated 20,000 to 25,000 human genes.

Some human traits are controlled largely by a single gene. But most inheritable characteristics are influenced by a number of genes that interact in a complex fashion. Also, personal experiences and environmental factors combine with genetic influences to shape certain traits, including vulnerability to disease and characteristics such as intelligence, emotions, talents, and personality.

Human Genome

Human genes reside on 23 pairs of chromosomes found in the nucleus of every body cell except gamete cells. In each pair, one of the chromosomes is inherited from the mother and the other is passed down from the father. About 2 m (7 ft) of DNA is packaged into each chromosome. All of the genes carried on chromosomes form the human genome. A lesser amount of DNA can be found in mitochondria, cellular organelles responsible for creating the energy used in cell activities.

All but one of these 23 pairs are composed of chromosomes nearly identical in shape. Each of these 22 chromosome pairs, known as autosomes, contains the same genes (although they likely carry different alleles). The autosome pairs vary considerably in length, and scientists number them according to their relative size: Pair number 1 is the longest pair and pair number 22 is the shortest.

Rounding out the human genome is the 23rd pair of chromosomes, known as the sex chromosomes, which determine the sex of an individual. Females inherit two X chromosomes, a matched pair carrying the same genes. One X chromosome is inherited from the mother and one X chromosome is inherited from the father. Males inherit an X chromosome from their mother and a Y chromosome from their father. The Y chromosome is shorter than the X chromosome and bears far fewer genes. One gene on the Y chromosome causes an embryo to develop as a male rather than a female.

Humans produce gamete cells for sexual reproduction. These gametes contain a haploid number of chromosomes—23 chromosomes instead of the full complement of 46. Female gamete cells mature into eggs, with each egg containing chromosomes 1 through 22 and an X chromosome. Males produce gametes that mature into sperm, and each sperm cell has a single set of chromosomes 1 through 22 and either an X or a Y chromosome. During fertilization, an egg that joins with a sperm containing a Y chromosome develops into a male, and an egg fertilized by a sperm containing an X chromosome develops into a female.

Human Genetic Disorders

Thousands of inherited diseases caused by altered genes and chromosomal abnormalities affect humans. These disorders cause problems such as physical deformities, metabolic dysfunction, and developmental problems. Medical surveys indicate that roughly 1 per cent of newborns in the United States have a single-gene defect. As many as 1 baby in 200 is born with a chromosomal abnormality serious enough to produce physical defects or mental retardation.

It is misleading to say that a person "inherits the gene" for a disease, since humans are born with the same number and types of genes. We inherit allele forms of specific genes, and these alleles may be defective. Most of the known inherited genetic disorders are caused by the mutation of a single gene, resulting in alleles that produce disease. These defects often produce disturbances in the body's biochemical processes, such as inhibiting the action of an important enzyme or stimulating the overproduction of a harmful substance. Frequently the consequences of such problems can cause severe disability or be fatal.

Many single-gene disorders follow Mendelian patterns of inheritance. A mother and father each pass an allele for a specific gene on to a child. If one of the alleles is defective and causes disease, the child will develop the disease according to a dominant-recessive pattern of inheritance. For example, cystic fibrosis (CF), a metabolic disorder that causes a progressive loss of lung function, is caused by a mutation in the recessive allele of a gene responsible for regulating salt content in the lungs. The recessive allele is unable to direct the production of a key protein, resulting in a salt imbalance that causes thick, suffocating mucus to build up in the lungs. If a baby inherits the defective allele from just one parent, no disease results. But the infant who inherits the defective allele from both parents will be born with the disease.

In other cases, a single dominant allele causes genetic disease. Huntington's disease, a condition characterized by involuntary movements, dementia, and eventually death, is caused by the inheritance of a pair of alleles in which a defective allele dominates the normal allele for the gene. An affected parent has a 50 per cent chance of passing the defective allele to a child. A child who inherits the dominant defective allele from just one parent will develop the disease.

Other inherited genetic diseases are caused by defects in the genes found on the X chromosome. Hemophilia, the inability of the blood to clot and heal a wound, is caused by a defect in an allele located on the X chromosome that helps produce proteins involved in the clotting process. Women who inherit this defective allele usually have the normal allele on their second X chromosome, which produces enough of these clotting proteins for the body to remain healthy. Women who inherit this faulty allele have a 50 per cent chance of passing the defective allele on to their children. Males who inherit this defective allele do not have a normal version of the allele on their Y chromosome and so cannot produce clotting proteins to heal wounds. Hemophiliacs are almost always males who have inherited an X chromosome with the faulty allele from their mother.

Other genetic disorders arise due to the inheritance of an abnormal number of chromosomes or a defective chromosome structure. These chromosomal abnormalities have a devastating impact: Many fetuses with such defects, particularly those with missing chromosomes, will die prenatally, resulting in miscarriage (spontaneous abortion). In other cases, newborns with chromosomal abnormalities suffer from physical problems or varying degrees of mental retardation. Down syndrome occurs when an individual's cells carry an extra copy of chromosome 21. People born with this condition have characteristic facial features, short stature, severe developmental disabilities, and a shortened life expectancy.

GENETICS AND CANCER

Cancer is a common name for many diseases that affect different body tissues, including the skin and the liver. All cancers involve alterations in genes that control cell division. These alterations cause cells to replicate abnormally and form tumors. Cancers generally arise from mutations that occur directly in the somatic cells, any cells of the body with the exception of the gametes (sperm and egg cells). Since the genetic mutations have not occurred in gametes, the mutations are not inherited by the next generation.

While cancer is not a traditional inherited genetic disorder, scientists have determined that a genetic component plays a strong role in the development of the disease. Geneticists have identified many different genes with certain alleles that appear to increase an individual's susceptibility to cancer. A notable example involves two genes linked to breast cancer. Researchers estimate that more than half of the women with a family history of breast cancer who inherit mutated alleles of these two genes, known as BRCA1 and BRCA2, will develop breast cancer by the age of 70.

In contrast, women who lack either of the mutated alleles have only a 13 per cent chance of developing the disease. For many cancers, researchers believe that mutations in several different genes must accumulate before cancer develops. As a person ages, errors in DNA replication may occur during cell division, or cells may be damaged by exposure to certain environmental factors, including cigarette smoke, radiation, and chemical pollutants. As a result, an accumulation of mutations may develop in two types of genes: tumor suppressor genes and oncogenes. Tumor suppressor genes normally function to halt cell division, while oncogenes function to activate cell division. A mutation in either type of gene can stimulate nonstop cell division. These types of defects have been linked to some cases of leukemia as well as to cancers of the ovaries, lungs, colon, and other organs.

Genetics and Aging

A growing area of study focuses on the link between aging and genetics. Scientists have determined that structures called telomeres, long, repetitive sequences of nucleotides at the end of chromosomes, affect the aging process. Each time a cell divides, telomeres become shorter. When the structures shorten to a certain length, the process of cell division terminates.

The cells of these chromosomes continue to live, but they never divide again. Laboratory tests suggest that an enzyme produced in gamete cells of the human body, called telomerase, can maintain telomere length in human cells, enabling them to continue dividing, perhaps indefinitely. Scientists hoping to lasso the life-extending properties of telomerase have been confounded by research indicating that telomerase is also active in rapidly dividing cancer cells. Before telomerase can be used to slow or halt aging, scientists must learn how to manipulate the enzyme so that it does not promote cancer growth.

Genes and Behaviour

Scientists actively explore the links between genes and behaviour to determine both the patterns and the limits of genetic influence. Such studies continue to be controversial because behaviour or mental processes can be difficult to measure objectively. Furthermore, many behavioral traits, both normal and abnormal, are complex, influenced by many genes as well as by personal experiences.

Studies of the possible genetic components of psychiatric disorders have yielded mixed results. Geneticists have identified at least two genes linked to schizophrenia, a condition characterized by hallucinations, delusion, paranoia, and other symptoms. Other studies that reported the discovery of genes that influence bipolar disorder (also known as manic-depressive illness) and alcoholism have been reversed or questioned. Though attempts to identify genes linked to these disorders have been flawed, scientists have little doubt that the conditions do have a genetic component.

Scientists have established links between genes and certain antisocial or violent behaviors. For instance, researchers have identified a gene on the X chromosome that has been tied to extremely violent behaviour in men. They identified the gene in members of several families with a multigenerational history of violent, criminal behaviour. They identified gene codes for monoamine oxidase inhibitor (MAO), an enzyme that helps nerve cells in the brain communicate with each other. Males in an affected family who inherit a defective allele for MAO do not produce enough of the enzyme. As a result, low levels of MAO change the activity of certain brain nerve cells, possibly contributing to socially unacceptable behaviour.

While research suggests that men who have a defective allele for the MAO gene are more prone to aggressive behaviour, experts cite numerous reasons for concern and doubt. Few scientists believe that a single gene could have a leading role in influencing complex behaviors. Others charge that these kinds of investigations promote an unreasonably simplistic view of genetic determinism, in which genes can be blamed for certain behaviors. Critics note that studies have identified many men who carried the defective allele for MAO production and never committed a violent act. Clearly, nongenetic influences—as varied as an individual's family life, work circumstances, attitudes, diet, and emotional state—affect complex behaviors.

IDENTIFYING GENETIC DISORDERS

Health-care professionals who specialize in genetic disorders use a variety of methods to identify inherited conditions. Analysis of a family medical history, known as pedigree analysis, is used to track the transmission of a condition through generations. Blood tests that identify specific DNA sequences can reveal carriers of a disease-causing gene who have no symptoms of the disease.

Geneticists collect a person's medical family history to trace the inheritance of a genetic trait among multiple generations. The information is placed in a pedigree, which resembles a traditional multigenerational family tree but includes information about individuals who were diagnosed with a particular disorder or who suffered from certain medical symptoms. A pedigree can help researchers recognize diseases that express themselves in dominant or recessive alleles. Dominant disorders affect every generation. Recessive disorders may cluster in a single generation, reflecting when two parents who both carry a recessive allele for a disease have one or more children who develop the disease. A pedigree can also identify diseases that show X-linked inheritance.

Pedigree analysis can be useful when combined with certain genetic tests. A blood sample taken from a person who is at risk for a genetic disorder can be compared with a DNA sequence known to cause the disorder in question. Other genetic tests can reveal if a person has extra chromosomes, missing chromosomes, or chromosomes that have attached to one another in unusual ways. In some cases, these chromosomal abnormalities may produce genetic disorders in children or they may affect a person's ability to conceive a child. Genetic testing can also identify disorders in a fetus, enabling parents to learn early in a pregnancy if a fetus will likely be born with health problems or develop them later in life.

Presymptomatic testing can identify DNA abnormalities in a person before health problems develop. In the case of certain inherited heart conditions, for example, these tests enable a person to make healthy lifestyle changes or take other preventative measures, such as medications, to lower the risk of illness or death.

Medical genetic testing raises challenging issues because such tests typically provide statistical possibilities rather than a definite prediction of whether a person will develop a given genetic disease. A test result may indicate, for example, that a person has a 75 per cent risk of developing colon cancer by the age of 65. Such results enable a physician to perform appropriate screening tests on the at-risk person in order to identify the disease at its earliest stages, when it is most treatable. At the same time, however, physicians must decide at what age the person's screening should begin and whether the benefits of early screening are worth the drawbacks of frequent screening. These drawbacks include expense, patient anxiety and discomfort, and exposure to radioactivity or other harmful substances used in testing. Different problems are posed by genetic screening tests that diagnose conditions for which no preventive measures exist, such as Alzheimer's disease, a progressive brain disorder that causes the loss of mental function. People may find it devastating to learn that they are at risk for a deadly disease that cannot be prevented by medical measures or lifestyle choices.

GENE THERAPY

A recent development in genetic technology known as gene therapy focuses on curing inherited disorders. In experiments using gene therapy, researchers have replaced defective genes with normal alleles, inactivated a mutated gene, or inserted a normal form of a gene into a chromosome. The earliest success in human gene therapy involved the treatment of infants who cannot produce adenosine deaminase (ADA), an enzyme important to normal function of the immune system. Scientists have successfully inserted the normal allele for the gene that codes for the enzyme into cells in ADA-deficient children. Preliminary evidence indicates that this gene therapy leads to better immune function in recipients. Researchers are also exploring gene therapy's potential to help treat people with many other conditions, including certain cancers, hemophilia, heart disease, and cystic fibrosis.

Although the United States Food and Drug Administration (FDA) has approved more than 400 clinical trials in gene therapy, this method of treating disease remains far from an unqualified medical success. Treatments usually produce some improvement in the underlying condition, but not enough to consider the therapy suitable for large-scale use. The death of a patient involved in a gene therapy experiment in 1999 caused the National Institutes of Health (NIH), a federal agency that monitors gene therapy studies, to reevaluate the safety and effectiveness of gene therapy clinical trials.

HUMAN GENOME PROJECT

The Human Genome Project is the most ambitious project in the history of biology. The program's challenging goal was to identify and sequence all of the DNA in human chromosomes. The project was initiated in 1990 in the United States with government funding, and it rapidly grew into an international consortium of academic centers and drug companies in China, France, Germany, Japan, the United Kingdom, and the United States. The consortium initially hoped to reach its goal by the year 2005.

In 1998 Celera Genomics, a privately funded biotechnology firm, announced that it would sequence the human genome by the year 2000 using different sequencing strategies than those used by the public consortium. This announcement triggered a heated race between Celera Genomics and the public consortium to complete the genome project. In June 2000 both teams declared victory when they jointly announced that they had separately completed a rough draft of the genome. The two teams published their findings simultaneously, although in two different journals, in February 2001.

The draft provided a basic outline of 90 per cent of the human genome. Scientists from the public consortium completed the final sequencing of the human genome in April 2003, two years earlier than planned.

The completed human genome has provided scientists with a detailed blueprint of our complex genetic code. Large computer databases of genetic information enable scientists to look for patterns and relationships among the actions of different genes. Among the findings about the human genome was that the number of genes in the human genome is much lower than was predicted—only about 20,000 to 25,000 genes compared to the expected 100,000 genes. This number is a little more than twice the number of genes found in the fruit fly.

Scientists are now turning their attention to studying how the relatively low number of genes in the human genome can produce the complex structures found in humans. Scientists have long known that a single gene produces a single protein and that this single protein subsequently may be processed into several different proteins. In a new science known as proteomics, scientists seek to identify and understand the function of all the proteins in the human body. They theorize that there may be many more proteins than there are genes—that is, more than 25,000. Among other advances, the database of proteins derived from proteomics is expected to help scientists better understand the regulation of gene expression in the body and how it leads to the complexity of cellular structures and functions. In addition, proteomics may lead to the development of breakthrough drugs for a variety of genetic disorders.

GENES AND OUR WORLD

Breakthroughs in decoding and manipulating the genetic information stored in DNA promise a world of benefits. These scientific advances already help to diagnose and treat disease, develop new medicines, bring criminals to justice, improve our food supply, and clean up the environment. At the same time, however, genetic technologies also present society with the potential for new and serious social or environmental problems. Many of the developments that worry critics of genetic technologies remain on the horizon, but the debate over their inevitable arrival is already in full swing.

Genes and the Environment

Humans have tampered with the genetic composition of other organisms for thousands of years. Most of this manipulation has been decidedly low-tech: domestication of animals and selective breeding of desirable food crops. The development and use of genetic engineering techniques has accelerated the pace at which humans can alter nature, creating some products that have unquestionable benefits and others that raise serious concerns. Some scientists fear that genetic engineering techniques will damage genetic diversity.

Domestication and selective breeding, which aim to produce many similar organisms with particular desirable traits, reduce the natural variation of genes within a species. Genetic engineering techniques take this effect a giant step further. Many crop species—including wheat, corn, tomatoes, and strawberries—have been manipulated to maximize yield, appearance, resistance to pests and chemicals, hardiness, and other commercially valuable traits. Once attractive varieties have been developed, agricultural techniques favour wide-scale planting of such genetically similar or identical stocks. The impact of this practice on biodiversity can be ominous because older plant varieties that carry diverse and useful alleles may be lost forever. The forfeited genetic material could leave a species vulnerable to annihilation from a single factor in the environment, such as an insect pest or an infectious virus.

Fortunately, efforts to combat decreasing biodiversity are under way. Farmers, gardeners, government agencies, and other interested parties have collaborated to create seed banks to maintain genetic diversity. These banks catalogue, store, and distribute the seeds of rare or endangered plants, enabling gardeners and farmers to continue cultivating rare plant varieties so that the unique genetic makeup of these plants does not disappear. In the same vein, zoos and other institutions breed endangered species of animals that may no longer be able to survive in their native habitats. Other animal programmes seek to preserve or enhance the genetic variation within certain endangered animal populations. For example, all cheetahs are almost identical genetically, most likely due to their near extinction about 12,000 years ago. Inbreeding among the few remaining individuals has resulted in a loss of genetic diversity in modern cheetahs that may have affected the cheetah's immune system, leaving the animal vulnerable to disease. Scientists hope to use genetic engineering techniques to introduce new genes into the cheetah population to increase the genetic diversity of the species.

The broad use of genetic engineering techniques in agriculture has raised other concerns beyond issues of biodiversity. From a consumer point of view, for instance, new technologies may have compromised a food's taste or nutritional value in exchange for a plant or animal that can be grown faster or at less cost. In addition, some critics question the safety of genetically engineered foods. They fear that plants or animals that have received new genes will produce proteins that would not be present in nonmanipulated organisms. Such changes could have a serious effect: causing allergies or toxicity in humans who eat these foods, for example, or disrupting a plant's production of key nutrients. Though there has been much speculation about the potential health risks of genetically engineered foods, rigorous scientific investigation into the effects of these foods on humans is just beginning. The potential environmental impact of genetically engineered agriculture is equally controversial. Critics fear that transgenic organisms, which contain DNA from other species, could give rise to populations of genetically altered life forms that could cause disease, displace native species, or otherwise harm delicate ecosystems. Consider the example of a genetically engineered form of oilseed rape, the plant that yields canola oil. The altered form, which has been grown commercially in the United States since 1993, contains inserted genes that increase its resistance to herbicides (weed-killing chemicals). The altered plants can grow unabated even when a farmer sprays a field with enough herbicide to kill troublesome weeds. Yet a curious problem has emerged: The transgenic rape can interbreed with weedy relatives that grow nearby. Some scientists fear that this interbreeding could create weed varieties that also have a genetic resistance to known herbicides. The biological and economic impact of such a development could be enormous.

Genes and Society

Advances in genetic technologies allow scientists to take an unprecedented glimpse into the genetic makeup of every person. The information derived from this testing can serve many valuable purposes: It can save lives, assist couples trying to decide whether or not to have children, and help law-enforcement officials solve a crime. Yet breakthroughs in genetic testing also raise some troubling social concerns about privacy and discrimination. For example, if an individual's genetic information becomes widely available, it could give health insurers cause to deny coverage to people with certain risk factors or encourage employers to reject certain high-risk job applicants. Furthermore, many genetically linked problems are more common among certain racial and ethnic groups—for example, the BRCA1 breast cancer allele is more common in Ashkenazi Jews, and the blood disorder sickle-cell anemia is more prevalent among blacks of African ancestry. Many minority groups fear that the expansion of genetic testing could create whole new avenues of discrimination.

Of particular concern are genetic tests that shed light on traits such as personality, intelligence, and mental health or potential abilities. Genetic tests that indicate a person is unlikely to get along with other people could be used to limit a person's professional advancement. In other cases, tests that identify a genetic risk of heart failure could discourage a person from competing in sports.

New technologies that allow the manipulation of genes have raised even more disturbing possibilities. Gene therapy advances, which allow scientists to replace defective genes with normal alleles, give people with typically fatal diseases new hope for healthy lives. To date, gene therapy has focused on manipulating the genetic material in body cells other than gametes, so the changes will not be passed on to future generations. However, the application of gene therapy techniques to gametes—the cells involved in reproduction—seems inevitable. Such manipulation might help prevent the transmission of disease from one generation to another, but it could also produce unforeseen problems with long-lasting consequences. For instance, many people worry that new genetic techniques could be used to alter or encourage traits now viewed as part of normal human variability, such as shortness or baldness. At various times in the past century, people have advocated efforts to improve the human condition by promoting the perpetuation of certain genes. This concept, known as eugenics, typically involves encouraging people with "positive" genes to reproduce and discouraging those with "inferior" genes from having offspring. Many people fear that new genetic technologies used to manipulate the human genome could give people previously unattainable methods to resort to extreme forms of eugenics.

Advances in genetic technologies have turned some genes into valuable commercial commodities, spawning a host of controversial questions. Who owns a genetically altered organism or the genes it contains? Is it right to patent the use of a naturally occurring gene? Some people feel that genetic material should not be owned or used for profit. Costa Rica has enacted laws to prevent foreign companies from patenting and then profiting from genes of native Costa Rican plant and animal species. Balancing the need to limit patents on genes are concerns that the profit motive of companies must be protected to maintain incentives to make new discoveries for medical products. The citizens of Iceland, for example, are cooperating with a biotechnology company in a study of the genetic makeup of the Icelandic people. The information compiled will be used to learn about genetic diseases.

7

Tissue Science Engineering and Technology

Tissue engineering is the use of a combination of cells, engineering and materials methods, and suitable biochemical and physio-chemical factors to improve or replace biological functions. While it was once categorized as a sub-field of bio materials, having grown in scope and importance it can be considered as a field in its own right.

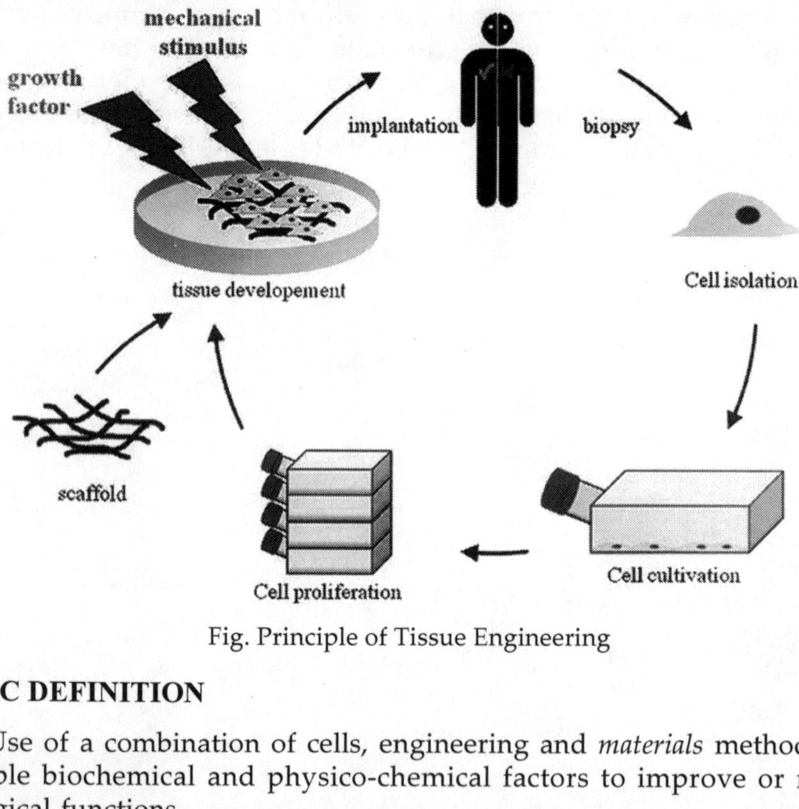

Fig. Principle of Tissue Engineering

IUPAC DEFINITION

Use of a combination of cells, engineering and *materials* methods, and suitable biochemical and physico-chemical factors to improve or replace biological functions.

Note: While most definitions of tissue engineering cover a broad range of applications, in practice the term is closely associated with applications that repair or replace portions of or whole tissues (*i.e.*, bone, cartilage, blood vessels, bladder, skin, etc.).

Tissue Engineering is an emerging multidisciplinary field involving biology, medicine, and engineering that is likely to revolutionize the ways we improve the health and quality of life for millions of people by restoring, maintaining, or enhancing tissue and organ function. Tissue engineering emerged as organ transplantation is limited by the number of available donors and high cost process, leaving thousands of people each year on the transplant waiting lists in the United States alone. Many die before an organ donor becomes available.

Tissue engineering aims at developing functional substitutes for damaged tissues and organs, a process that involves the use of a combination of cells, engineering and material methods, including suitable biochemical and chemical factors to improve or replace biological functions. Rather than simply introducing cells into a diseased area to repopulate a defect and/or restore function, in tissue engineering the cells are often seeded in or onto biomaterials (scaffolds) before transplantation.

Until recently, it was believed that the macroporous features of scaffolds used in tissue engineering mimicked the dimension scale of the extracellular matrix (ECM), and that the matrix itself (natural or artificial) only served as a support for the cells; morphogenesis was controlled passively by defining tissue boundaries. Emphasis was placed on critical engineering and material issues, such as improving mass transfer into the core of the cell constructs and designing biocompatible and biodegradable scaffolds with mechanical properties suitable for engineering various tissues. As the field evolved, attention focused on the biology of the scaffolds and how they affect various cell types. Tissue engineers had recognized that some of the widely used scaffolds do not fairly recapitulate the cell microenvironment and that the ECM is a dynamic and hierarchically organized nanocomposite that regulates essential cellular functions such as:

- Morphogenesis,
- Differentiation
- Proliferation
- Adhesion
- Migration

Nanotechnological tools for tissue engineering may help design advanced nanocomposite scaffolds that can better mimic the ECM and eventually assemble more complex and larger functional tissues. In order to generate a functional tissue, effective organization of cells in the tissue is required with similar morphology and physiology of the parental tissue. Morphogenesis in the three-dimensional (3D) scaffold should occur in a similar way to natural organ development. The cells reorganize owing to interaction with the ECM on the basis of:

- Topography,
- Mechanical properties (such as matrix stiffness, elasticity and viscosity)
- Concentration gradients of immobilized growth factors
- ECM molecules.

Recently, Ott and co-workers (4) reported a study emphasizing the importance of the ECM structure in guiding the seeded cells and promoting morphogenesis. Rat hearts were decellularized by perfusion of detergents to preserve the underlying ECM and then reseeded with cardiac and endothelial cells (4). The cells migrated and self-organized in their natural location in the matrix and by day 8, under physiological load and electrical stimulation, the constructs were able to generate pump function (4). The importance of the ECM was shown for:

- Heart
- Lung
- Arteries
- Liver
- Bone
- Nerve

So why is the Extracellular Martix (ECM) so important?

The ECM is composed of an intricate interweaving of protein fibres such as fibrillar collagens and elastins, ranging from 10 to several hundreds of nanometres. The mesh is covered with nanoscale adhesive proteins such as laminin and fibronectin that provide specific binding sites for cell adhesion (interacting with integrins, cadherins and so forth) and have been shown to regulate important cell behaviours such as growth, shape, migration and differentiation. Polysaccharides such as hyaluronic acid and heparan sulphate fill the interstitial space between the fibres and act as a compression buffer against the stress placed on the ECM or serve as a growth factor depot.

Scaffold design considerations

The ECMs of various tissues in the body differ in the composition and spatial organization of the collagens, elastins, proteoglycans and adhesion molecules, to maintain specific tissue morphologies and organ specific shape and function, and to supply specific instructive cues. Therefore, the design considerations for scaffolds should vary according to the desired engineered tissue. For example, the biochemical, electrical and mechanical functions of the heart are uniquely dependent on their biological nanostructures. The heart's 3D ECM network is composed of an intricate, micro- and nanoscale interweaving pattern of fibrillar collagen and elastin bundles that form a dense, elastic network with proteoglycans and with adhesive and non-adhesive molecules. In this defined mesh, the cardiomyocytes are forced to couple mechanically to each other, to form elongated and aligned cell bundles that interact with each other or with neighbouring capillaries and nerves.

Post-isolation cells lose their ultrastructural elongated morphology and their interaction with their surroundings, and they adopt a random distribution on the flat surface of the scaffold, which compromises many of their physiological properties. Therefore, the structure and support of the ECM is crucial. See Figure 2.

Limitations of the ECM:

- Weak mechanical properties
- Lack of electrical conductivity
- Absence of adhesive and micoenvironment- defining moieties
- Inability of cells to self-assemble to 3D tissue structure.

The Heart for example requires more than alignment and mechanical support (Boyang Zhang, Ref 5)

1. Cell responses to micro- and nanopatterned topographical cues
2. Cell responses to patterned biochemical cues
3. Controlled 3D scaffolds
4. Patterned tissue vascularization
5. Electromechanical regulation (conductivity). of tissue assembly and function

Nanostructures can be used to record the electronic signals that are transmitted through cells such as neurons and cardiomyocytes. One way to record these signals is by lithographically defining nanostructures as field-effect transistors, which are sensitive to local electric field changes. In particular, silicon nanowire transistors are useful for measuring extracellular signals because they exhibit particularly exquisite field-effect sensitivity compared with conventional, planar devices; they are just tens of nanometres in diameter and can therefore interface with cells and tissue at a subcellular level; and they show nanotopographic features that encourage tight interfaces with biological systems.

HISTORY OF TISSUE ENGINEERING

Providing a history of tissue engineering it is inevitable to return to the basics, starting in 1665, when Hooke (1635-1703) discovered small holes in cross-sections, which he called cells and described in his book *Micrographia*. In 1805, Okenstated, "All life is based on individual cells."

In 1838-1839, Schleiden and Schwann formulated the so-called "Cell Theory" based on their microscopic findings. This theory summarized their findings as:

- The cell is the unit of structure, physiology, and organization in living things
- The cell retains a dual existence as a distinct entity and a building block in the construction of organisms

- Cells form by free-cell formation, similar to the formation of crystals (spontaneous generation).

In 1858, Virchow (1821-1902) described his ideas about cell formation with the now famous words, "Omnis cellula e cellula... " saying that cells arise from pre-existing cells, confuting Schleiden and Schwann's idea of spontaneous cell generation. Virchow presented his ideas about regeneration stating that tissue regeneration is dependent on cell proliferation. This led research to focus attention toward the more fundamental cellular level. Thiersch attempted to grow skin cells into granulating wounds. In doing this, he discovered the important influence of granulation tissue on wound healing in 1874.

Loeb first reported the idea of growing cells outside the human body in 1897. From that point on many researchers experimented with growing cells *in vitro*. While experimenting with different media, survival was estalished, despite no growth. Harrison (1870-1959) was the first to grow frog ectodermal cells *in vitro* in 1907, thus developing the first neuronal tissue culture line. This was followed in 1912 by Carrel who was able to grow pieces of chick embryo in various media, which he initially maintained for 85 days, and subsequently for years. Later, interest arose in growing cells instead of complete tissues. In 1916, Rous and Jones discovered that trypsin is capable of degrading matrix proteins, thus separating cells. Trends changed toward expansion of cell types throughout the 1940s and soon after, much research was performed, which led to the ability to grow tissue-specific cell lines *in vitro*. Enders contributed greatly to the use of human embryonic cells in 1952. In 1998, the development of stem cell lines formed the basis of modern tissue engineering. Research since 1998 has focused on identifying the differential capacities of embryonic and adult stem cells and how to influence and accompany this differentiating pathway.Aside from this, research focuses on the problems associated with expansion of autologous and allogeneic differentiated cells, as a large number of cells are needed to generate a relatively small volume of tissue. Furthermore, fundamental research is continually being performed to gain insight into the cellular interactions that are of great importance in tissue engineering.

TECHNIQUE

Ingredients

Four components are essential for tissue engineering: (stem) cells, a matrix or scaffold, a bioreactor, and cytokines. In tissue engineering, a scaffold becomes embedded with living cells and specific regulatory cytokines, and is placed into a bioreactor. Ideally, a suitable biochemical and biomechanical microenvironment is created and cell multiplication fills the scaffold with tissue and allows the cells to grow into the correct shape. When implanted into the body, the seeded scaffold becomes integrated concomitantly supporting and directing cell proliferation. As the cells proliferate the scaffold slowly biodegrades, gradually allowing blood vessels and host cytokines to make contact with the cells. Through this process, the scaffold further biodegrades while the cells proliferate and differentiate into the desired tissue. Finally, the scaffold completely dissolves and the formed tissue starts functioning in its new surrounding.

Stem cells

Because every tissue consists of a variety of cell types, it is desirable to use common progenitor cells that have the embryological ability to differentiate into multiple cell types. Pluripotent cells are more often referred to as stem cells. Stem cells are also able to recreate themselves indefinitely and maintain their own population. Stem cells can be isolated from an embryo, fetus, or an adult.

The use of embryonic stem cells in medicine is controversial. Many religious and ethical groups express concerns with their use. A benefit of adult stem cells (ASC) use is that it does not require destruction of embryos. Moreover, in regards to potential skin substitutes, recent research has shown that epidermal adult stem cells have a greater plasticity than was initially expected. In an animal model, adult epidermal stem cells were able to differentiate into different functional cell types. One additional advantage of ASC (and especially epithelial) use is the ease of collection from organisms.

Extracellular Matrix and Scaffold

A matrix or culture is needed to facilitate cell growth. *In-vivo*cells are arranged in the extracellular matrix (ECM) compromising a complex structural entity surrounding and supporting cells.The ECM is necessary in order for certain cells to carry out their functions. The ECM has 3 major components: structural proteins (mainly collagen); proteoglycans and hyaluronan; and specialized multiadhesive proteins. Every tissue contains itsown type of ECM specialized for its particular function. The amount of specific components varies according to the function of the tissue.As a result of the minimal availability of extracellular space in the epidermis, cells are tightly bound to the basal lamina, which is a thin matrix overlying the loose connective tissue in the dermis.

Proteoglycans are macromolecules consisting of a core protein attached to several polysaccharides (glycosaminoglycans). Proteoglycans are able to bind cells to the matrix and bind growth factors (eg, fibroblast growth factor [FGF], transforming growth factor beta [TGF-β] to glycosaminoglycans, thus preventing extracellular protease degradation of growth factors. Organization of all ECM components depends on the presence of binding proteins in the ECM. Fibronectins are another class of matrix proteins that can bind to integrins and play an important role in attaching cells to the ECM. In tissue engineering, much research focuses on developing artificial matrices or scaffolds. Scaffolds can be permanent or temporary depending on the application. Permanent scaffolds are not biodegradable and remain incorporated in the body. These scaffolds are necessary in situations where continuous strength is needed to allow tissue to retain its shape. Today, mostly temporary and biodegradable scaffolds are used ultimately leaving only the engineered tissue *in situ*.

Research has shown that combining collagen and hyaluronic acids in scaffolds, in addition to using hyaluronic acid matrices, has been successful in tissueengineering of cartilage and dermal tissue regeneration.

Bioreactor

A bioreactor can be defined as any apparatus that attempts to mimic and reproduce physiological conditions in order to maintain and encourage cell culture for tissue regeneration.Cell-culture parameters such as temperature, pH, biochemical gradients, and mechanical stresses should be continuously controlled during the maturation period.

Bioreactors have already improved the processing and the final results of skin regeneration. It is essential that bioreactors are designed and fabricated following specifications that differ from tissue to tissue.

Cytokines

Cytokines are polypeptides that can bind to cell-specific receptors present in cell membranes of target cells triggering cell-specific responses. These polypeptides usually exist as inactive precursors that need to be cleaved and bound to the ECM to become active. By binding receptors cytokines can exert a signal transduction cascade via second messengers leading to changes in gene expression mediating specific cellular responses. Responses of binding a cytokine can include altering the expression of membrane proteins, secretion of effector molecules, and proliferation. Cytokines are not cell type specific—a single cytokine can act on multiple cell types and different cell types can secrete the same cytokine, despite differing responses between different cell types.

Epidermal growth factor (EGF), fibroblast growth factor (FGF), platelet-derived growth factor (PDGF), and TGF-β are the 4 most important growth factors in wound healing. EGF stimulates growth and migration of keratinocytes *in vitro* and epidermal regeneration *in vivo*. Epidermal growth factor has already been applied to delivery systems leading to the enhancement of epidermal regeneration in partial-thickness wounds and second- degree burns. Once applied to a hydrogel, significantly faster wound healing was observed. Fibroblast growth factor (FGF) acts on fibroblasts. Of the 20 FGF family members,FGF-β is especially known as a potent stimulator of endothelial cells that induces angiogenesis. Platelet-derived growth factor (PDGF) is also important in wound healing as it attracts various cells to the lesion, including fibroblasts, smooth muscle cells, neutrophils, and macrophages. It also stimulates secretion of growth factors by macrophages and matrix products, such as fibronectin, hyaluronan, collagens, and proteoglycans. PDGF plays an important role during the remodeling phase by stimulating the fibroblast production of collagenases. Recombinant PDGF is therapeutically applied to chronic diabetic ulcers enhancing wound healing.

The transforming growth factor-β (TGF-β) super family consists of more than 40 members with a similar structure. Almost all cells in the human body secrete inactive TGF-β. In order to bind its receptor,TGF-β must be activated. Binding to its receptor initiates a cascade mediated by activation of a protein kinase. Research has shown that incorporation of TGF-β in a collagen scaffold resulted in a faster epithelialization and contraction rates in full-thickness skin defects in rabbits. Research focuses on the development of delivery systems for TGF-β that can gradually release TGF-β.

TGF-β1 is a pleiotropic growth factor, which has reguvanlatory effects on many different cell types. For instance, it plays an important role in cell proliferation and differentiation, bone formation,angiogenesis, neuroprotection, and wound repair. However, long exposure to high doses of TGF-β1 results in fibroses and hypertrophic scars. Literature regarding cytokine manipulation of proliferative scars has shown that TGF-β2 may be involved in the development of tissue fibrosis.

In wound healing, TGF-β3 has been demonstrated to attenuate type 1 collagen synthesis and reduce scar tissue formation. Despite previous meritorious efforts to investigate the therapeutic potential of TGF-β3, its effective use is limited by a number of common shortcomings, such as a short half-life, *in-vivo* instability, and relatively inaccuracy of the delivery system.

WOUND HEALING AND TISSUE ENGINEERING

History

When considering wound healing historically, a trend can be seen throughout the centuries representing a shift from dry wound healing to wet wound healing. The oldest medical texts describing ancient medicine are written on papyrus that date from 1500 to 1000 BC and are based on much older initial documents dating from approximately 3000 BC. These documents indicate that in 3000 BC hemorrhages were cauterized to stop the bleeding. Around 2000 BC bandages began being used for approximating the wound edges. Later, around 1000 BC, Homer described 147 wounds in the Iliad. Throughout history different materials have been used for bandaging wounds. The indication of bandages shifted as Celsus in 25 AD described signs and symptoms of inflammation. Since then, bandages were used for prevention of the described symptoms and signs. In the Middle Ages, Paracelsus wrote of dressing wounds made with cotton and gauze. In the 19th century, Pasteur favored the use of dry dressings to help prevent infection. In 1962,Winter demonstrated that partial-thickness wounds reepithelialized more rapidly under occlusive dressings because occlusive dressings maintained a moist wound surface.This landmark study showed that a moist environment accelerated the re-epithelialization process. Numerous studies followed that demonstrated wound occlusion and moisture increased all phases of healing. Occlusive dressings are designed to keep the wound bed moist with exudate. The basic concept behind moist wound healing is that the presence of exudate in a wound will provide an environment that stimulates healing by delivering a range of cells and cytokines necessary for wound repair. With these findings, use of dry dressings became less popular, and lead to the development of occlusive dressings such as, hydrocolloids, hydrogels, semipermeable films, alginates, and foams.

Since 1992, research indicates that wet wound healing might also accelerate the healing process when compared to dry wound healing. Later, in 1995 it was shown that wet wound healing showed less subepidermal inflammatory cells when compared to moist wound healing represented by a hydrocolloid dressing. Wet wound healing uses a plastic transparent chamber filled with a medium that is attached to the wound.The chamber makes sure that the wound is constantly in contact with the medium and allows visible sight of the wound bed. Fluid chambers function as a quasi mini-incubator that make it possible to inject antibiotics and analgesics to the medium. With developments in tissue engineering emphasizing the importance of adding growth factors, wet wound healing is suitable for allowing cytokines and cells to be added to the medium.

Tissue-engineered Wound Healing Products

Tissue-engineered wound healing products can be classified as acellular or cellular (Table 1). Acellular products contain, as the name implies, no cells and consist of a matrix that functions by binding to the host, allowing matrix-cell interactions. Due to its porous nature, the matrix allows host cells to infiltrate.Today, a matrix can contain virus vectors or plasmids, which can transcribe and translate the in-built DNA leading to secretion of specific growth factors.

These growth factors carry out their specific functions by stimulating host cells to enhance wound healing. In specific situations it is even possible to manufacture matrices containing plasmids that can release hormones. The matrix contains ECM-proteins such as collagen, hyaluronic acid, and fibronectin, thus ensuring biocompatibility. Examples of these products are: E-Matrix(Encelle Inc, Greenville, NC), OASIS® (Healthpoint Ltd, Fort Worth,Tex), Integra® (Integra LifeSciences, Plainsboro, NJ), Permacol® (Tissue

Science Laboratories Inc,Andover, Md), Matriderm® (Dr. Suwelack Skin & Health Care AG, Germany), and EZDerm® (Brennen Medical, St. Paul, Minn). Restoring function after hand burns plays a major role in the restitution of quality of life.The grafted areas are of utmost importance for good hand function. Haslik et al evaluated the collagen-elastin matrix Matriderm as a possible alternative for 2-stage substitutes after hand burns.This matrix was evaluated in several trials by van Zuijlen et al and showed the sufficient survival of autografts on top of the dermal substitutes in a 1-step procedure. Haslik et al concluded that Matriderm is a promising dermal substitute for the treatment of severe hand burns, with an overall take rate of 97%.

Cellular products on the other hand do contain living cells, often fibroblasts and keratinocytes embedded in a collagen or polyglactin scaffold forming an epidermal layer skin substitute.Autologous cells are used in these products to minimize the risk of rejection. Autologous keratinocytes are derived from progenitor cells from dermal sheets in the outer root surrounding hair follicles or from epithelial cells obtained via a biopsy of the recipient's skin. A permanent autologous epidermal skin graft is applied, which functions as a reliable barrier and promotes the formation of granulation tissue. Examples of tissue-engineered epidermal substitutes are: Epicel (Genzyme Biosurgery, Cambridge, Mass), Laserskin® (Fidia Advanced Biopolymers, Abano Terme, Italy), Myskin™ (Celltran Ltd, Sheffield, UK), and EpiDex™ (Modex Therapeutics, Switzerland).

Laserskin autograft is an epidermal substitute. There are orderly arrays of laser-perforated microholes for the ingrowth and proliferation of keratinocyte. As keratinocytes are directly cultivated on Laserskin, the graft can easily be peeled off from the skin. Lam et al showed in an animal experiment that composite Laserskin grafts are good human skin substitutes in terms of durability, biocompatibility, high seeding efficacy for keratinocytes, high graft take rate, and low infection rate.

Aside from epidermal substitutes, tissue-engineered dermal substitutes have also been developed. The dermis is composed of loose connective tissue containing collagen and fibrils, anchoring the dermis to the epidermis. The upper layer of the dermis, the papillary layer, is a cell-rich layer containing fibroblasts and macrophages allowing dermis-epidermis interactions. These interactions trigger synthesis of ECM components and stimulation of differentiation and growth of keratinocytes. Inclusion of living fibroblasts leads to active release of cytokines. However, until now, only allogeneic tissue-engineered dermal wound products have been developed, and unfortunately always carry the risk of (chronic) graft rejection. Examples of these products include: Dermagraft® (Advanced BioHealing, La Jolla, Calif), Alloderm® (LifeCell Inc, Branchburg, NJ), TransCyte® (Advanced BioHealing, La Jolla, Calif), and ICX-SKN® (Intercytex, Cambridge, UK). Dermagraft is a cryopreserved human fibroblastderived dermal substitute. Fibroblasts incorporated with Dermagraft secreted VEGF, PDGF, insulin-like growth factor I (IGF-I),colony stimulating factors (CSF), interleukins (IL), tumor necrosis factor (TNF), and TGF-β. Alloderm is a dermal collagen matrix derived from banked human skin that is specially treated to remove most cellular components. Alloderm has been successvanfully used in the resurfacing of full-thickness burn wounds in combination with an ultra-thin autograft that replaces the epidermis.

TransCyte is a laboratory-grown extracellular matrix of allogeneic human dermal fibroblasts. Noordenbos et al reported on the safety and efficacy of TransCyte for treatment of partial-thickness burns. Wounds treated with TransCyte healed faster, with no infection and less hypertrophic scarring,compared to wounds treated with silver sulfadiazine. More recently, ICX-SKN, an autosynthesized human collagen-based extracellular matrix with human dermal fibroblasts vascularized during healing of acute surgical wounds and showed integration and persistence during the healing process. Currently, (allogeneic) bilayered products are available— eg, Apligraf® (Organogenesis, Canton, Mass) and OrCel® (OrCel International, New York, NY). Apligraf, a living bilayered skin substitute, is capable of regenerating tissue in response to an injury.This is apparent because of the good interaction between Apligraf and the wound, thus making it suitable for wide application. OrCel is a bilayered cellular matrix in which normal human allogeneic skin cells (epidermal keratinocytes and dermal fibroblasts) are cultured in 2 separate layers into a type I bovine collagen sponge.

Like Apligraf, OrCel is a bilayer dressing resembling normal skin. OrCel was developed as a tissue-engineered biological dressing. OrCel delivers ECM components and growth factors and creates an environment conducive to wound healing—it was never intended as an artificial skin for grafting.

TISSUE CULTURE

In many cases, creation of functional tissues and biological structures *in vitro* requires extensive culturing to promote survival, growth and inducement of functionality. In general, the basic requirements of cells must be maintained in culture, which include oxygen, pH, humidity, temperature, nutrients and osmotic pressure maintenance. Tissue engineered cultures also present additional problems in maintaining culture conditions. In standard cell culture, diffusion is often the sole means of nutrient and metabolite transport. However, as a culture becomes larger and more complex, such as the case with engineered organs and whole tissues, other mechanisms must be employed to maintain the culture, such as the creation of capillary networks within the tissue.

Another issue with tissue culture is introducing the proper factors or stimuli required to induce functionality. In many cases, simple maintenance culture is not sufficient. Growth factors, hormones, specific metabolites or nutrients, chemical and physical stimuli are sometimes required. For example, certain cells respond to changes in oxygen tension as part of their normal development, such as chondrocytes, which must adapt to low oxygen conditions or hypoxia during skeletal development. Others, such as endothelial cells, respond to shear stress from fluid flow, which is encountered in blood vessels. Mechanical stimuli, such as pressure pulses seem to be beneficial to all kind of cardiovascular tissue such as heart valves, blood vessels or pericardium.

Fig. Bioreactor for cultivation of vascular grafts

Bioreactors

A bioreactor in tissue engineering, as opposed to industrial bioreactors, is a device that attempts to simulate a physiological environment in order to promote cell or tissue growth in vivo. A physiological environment can consist of many different parameters such as temperature and oxygen or carbon dioxide concentration, but can extend to all kinds of biological, chemical or mechanical stimuli. Therefore, there are systems that may include the application of forces or stresses to the tissue or even of electrical current in two- or three-dimensional setups.

In academic and industry research facilities, it is typical for bioreactors to be developed to replicate the specific physiological environment of the tissue being grown (*e.g.*, flex and fluid shearing for heart tissue growth). Several general-use and application-specific bioreactors are also commercially available, and may provide static chemical stimulation or combination of chemical and mechanical stimulation.

Bioartificial Organs

An article printed in the New York Times "A First: Organs Tailor-Made With Body's Own Cells," it recognized the first procedure of regenerative medicine of an implantation of a "bioartificial" organ. In this case it was a windpipe, that was suggested by a doctor at the Karolinska Institute in Sweden. This and the artificial bladder is just the start to what scientists and doctors want to achieve. Goals of making more complex organs such as the heart, liver, and kidneys are just some of few. Back in 2011 a patient was going to die unless he received a new windpipe.

Dr. Macchiarini proposed it to his patient informing him that this procedure had only been tested on pigs. After slight hesitation and doubts he accepted. This procedure involved stem cells, but also a scaffold that could be used to make a whole artificial organ. The doctor extracted cells from the patients bone marrow and then with a synthetic scaffold they drip it "like basting a turkey" states this article in the New York Times. This continues until the human cells regenerate on their own and forms the organ needed. This kind of procedure helps eliminates some issues of a regular organ donated. This is "Because the donor's own cells are removed, this approach all but eliminates a major problem of transplants: the risk that foreign tissue will be rejected by the recipient." Dr. Macchiarini has more plans and goals however. He wishes to one day make organs without the synthetic scaffold which in the end is the main goal of most scientists and doctors working on tissue engineering.

NEED FOR A TISSUE-ENGINEERED VALVE

Some may argue that there is not much of a need for tissue-engineered valves, because conventional valve technology is very mature with well-described performance criteria. Indeed, the original motivation for the continued development of prosthetic valve technologies appears to have lessened considerably with the improvements in surgical approaches and hence surgical outcomes. In the early days of bioprosthetic valve development, durability and hence longevity, was the main motivating factor. The reason why durability was the main object of research and development is that prosthetic valves are meant to be implanted once and should last the life of the patient.

Historically, the main complication associated with the use of prosthetic valves has been operative mortality during prosthetic valve replacement surgery. Whereas the mortality during the first surgery—the surgery to replace the disease native valve—is often less than 1%, the second surgery is considerably more risky, with repeat operations having mortality as high as 20%, varying highly across institutions. With improvements in surgical technique and technologies, reoperative mortality has been reduced considerably during the past decade. Many surgeons now view surgical mortality during reoperation to be lower than the cumulative risk of thromboembolism associated with the use of mechanical valves (approximately 4% per patient-year) and now opt for repeated use of bioprosthetic valves in their patients. Indeed, the durability of the Edwards pericardial valve, perceived by many to be the most durable bioprosthesis, is close to 20 years and almost equivalent to the aortic valve homograft which, because of its excellent longevity, is considered by many to be the gold standard.

Because of the relatively good performance of current generation prosthetic valves, and the excellent quality of life they provide, both surgeons and valve manufacturers have become quite conservative. Surgeons are often reluctant to switch from a proven valve to a new device, and manufacturers are wary of introducing a new product that will have clinical performance that is worse than its current product. A number of clinical failures of new products have all but stifled innovation in conventional heart valve technologies. The failure of the Carbomedics Photofix-˜ valve (cusp abrasion and perforation), the Medtronic Parallel mechanical valve (thromboembolism), and the St Jude Silzone coating (tissue necrosis and perivalvular leak) have made the valve industry highly aware of the possibility that innovation can breed disaster. Most recently, Edwards Lifesciences suspended the clinical trial of its catheter-deployable valvebecause of problems with delivery and anchoring of the valve in the aorta of the patient.

In the world of conventional valve technologies, the bar for the adaptation of a new valve is very high. Current bioprosthetic valves have a lifespan of 15 to 20 years, highly predictable failure patterns that can be managed, and negligible early complications. A tissue-engineered valve will, therefore, need to compete in this arena, where change is slow and methodical, often taking a generation to establish. Indeed, the Edwards valve has increased its market share steadily over the past 20 years, primarily because of clinical reports on its very good long-term performance.

There may not be any immediate need for a tissue-engineered valve in the adult patient. Given the good service life that conventional valves offer, experimental use of new tissue-engineered valve concepts is questionable. The more realistic option for the use of tissue-engineered valves in the near future is in the pediatric population. The performance of the many surgical corrections for valvular defects is highly variable and depends on the age of the child. Allograft valves are difficult to obtain for children because they require the death of other children of similar size. Surgical reconstruction and monocusp valves tend to fibrose and contract early. In the child, there is a need for new materials and new approaches and thus an opportunity for tissue-engineered valves. Unfortunately, the market size for pediatric products is very small, less than 10% of the adult valve market, and thus not commercially viable. Accordingly, established heart valve manufacturers expend relatively few resources developing tissue-engineered valve technologies (neither Medtronic nor Edwards Lifesciences has a significant research program in heart valve tissue engineering; Medtronic and Edwards

Lifesciences, personal communications, 2005). Clinical use of tissue-engineered valves will thus most likely happen first in pediatric hospitals on an ad hoc basis and will augment the portfolio of surgical options currently available to treat these very challenging patients.

THREE APPROACHES TO VALVULAR TISSUE ENGINEERING

Perhaps the first examples of heart valve tissue engineering were the experiments with seeding cells on otherwise inert substrates. This approach has been relatively successful in vascular grafts but not at all for cardiac valves. Compliant synthetic materials have failed as valve leaflet substitutes, not because of thromboembolism but because of material fatigue, often related to microcracks, plasma protein insudation, and subsequent mineralization. Because of these fundamental material issues, glutaraldehyde-preserved biological matrixes have dominated the valve field. Because of the residual toxicity of glutaraldehyde, glutaraldehyde-treated valve tissues have remain cell free. For reasons not completely clear, they do not induce thromboembolism and remain an essentially passivated structured in the blood stream, which does not elicit any adverse reaction.

Perhaps the first example of heart valve tissue engineering came out of the University of Vienna in 1991, when Grimm et al presented their success in inducing endothelium to grow on glutaraldehyde-fixed bovine pericardium.Sustaining cells on the otherwise toxic pericardial tissue was made possible by inactivation of the aldehyde by crosslinking with L-glutamic acid before cell seeding. For unknown reasons, this technology was never implemented clinically and the last article on this topic was published in 1993.

The approaches that have been sustained by investigators around the world can be grouped into the following two areas: (1) decellularization of xenogenic tissues followed by cell seeding or direct implantation and (2) use of bioresorbable synthetic scaffolds. A third, less popular approach, involves fabrication of cell matrix constructs by way of polymerization and cell entrapment.

Acellular Matrix Xenograft

This is perhaps the oldest approach to mainstream valvular tissue engineering, the first reports being patents by Brendel and Duhamel of the University of Arizona, Tucson, filed in 1984, and by Klement et al from Toronto, filed in 1987. Both of these approaches claimed the much larger field of acellular matrix use for applications ranging from valves to vascular grafts to bone, teeth, ligament, and skin. Decellularization of heart valves has since been attempted all over the world. In the US and Canada, this approach was attempted by Vesely and Noseworthyand Wilson and colleagues very early on in this field and more recently by the thorough investigation of Hilbert et al. In Europe, this approach has been used more widely by Dohmen et al of Berlin, Germany, Steinhoff and colleagues of Rostok, Germany, Stock and colleagues of Jena, Germany, Haverich and colleagues from Hanover, Germany, Weigel et al from Vienna, Austria,Gittenberger-de Groot and colleagues from Leiden, The Netherlands, Fisher and colleagues of Leeds, UK, and Spina, Gerosa, and colleagues from Padua, Italy.In Asia, the approach has been adopted by Hong and colleagues and Ye et alfrom Shanghai, China, and by Wu et al from Beijing, China.

The rationale for this approach is the assumption that the antigenicity of xenogenic tissues originates in the cellular debris. Recall that porcine aortic valves need to be crosslinked with glutaraldehyde before implantation and that human allograft valves do not. Clearly, the immunohistocompatability mismatch between humans and pigs is far more severe than it is between unmatched human subjects. Pig cells express the gal-˜ epitope, whereas humans do not.

Indeed, human allograft valves are almost never matched to recipients, even though matching has been shown to have better long-term graft survival. Lack of matching is apparently not too detrimental because homograft valves can last 15 to 20 years. It has been theorized that perhaps the valve is in a privileged location, in a high flow environment where monocytes and other immune system cells cannot readily attach. With all of these issues in mind, suggesting that cell-extracted porcine valves can exist in human patients, without any appreciable antigenic response, may sound quite reasonable. This approach also assumes that these acellular matrixes will become repopulated with recipient cells, either before or immediately after implantation in the patient.

A fully repopulated matrix would thus become "invisible" to the host immune system, because it would be enveloped in an endothelial cell layer recognized as "self," and possibly remodeled by the invading cells, ultimately becoming self as the host cells lay down a new matrix in place of the degraded porcine matrix. This approach has been developed to the greatest extent by Goldstein et al of CryoLife Inc, a company well known for its homograft valve cryopreservation business. CryoLife eventually brought the technology to clinical use in Europe, unfortunately with disastrous results that are reviewed later in this article.

The typical approach to generating an acellular matrix tissue is to first break apart the cell membranes through lysis in hyper- and hypotonic solutions, followed by extraction with various detergents. The detergents used by most investigators include the anionic Sodium dodecyl sulfate, the zwitterionic CHAPS and CHAPSO, and the nonionic BigCHAP, Triton X-100, and Tween family of agents. The enzymes that have accompanied these detergent treatments have focused mainly on cleaving and removing the DNA that is part of the cellular debris. Because these enzymes can potentially degrade the useful matrix, enzyme inhibitors, such as trypsin inhibitor, can be used, although some have used trypsin-EDTA alone to decellularize the matrix. It is important to note that the agents used for cell extraction can be quite detrimental to the matrix: they can degrade or denature the matrix proteins or leave toxic residues or residual charge, any of which can detrimentally affect mechanical function or cellular response. The parameters that the many investigators in this field have varied involve mainly the sequence of steps, the specific detergents to use, and the time duration of the various soaking periods.

For example, the laboratory of Fisher in Leeds, UK, uses a series of baths in PBS and hypotonic buffer, along with trypsin and nucleases, to decellularize the tissue. The entire extraction procedure can last up to a week. Mechanical testing is used most of the time to determine the mechanical integrity of the processed tissues. Parameters such as burst pressure or failure strength of test trips is compared with unprocessed controls. In most cases, mechanical properties remain well preserved after cell extraction. Histological morphology, however, varies greatly from process to process, often showing a highly porous, locally collapsed microstructure, suggesting that there are features of the matrix that are not readily measurable using conventional techniques. The patents that describe the creation of these acellular matrixes are perhaps the largest collection in this field and number in the dozens, if not hundreds.

To seed or not to seed cells before implantation has become 1 of the important variables to consider in this approach. Perhaps the most disconcerting issues in this approach are reports that homograft valves not only fail to repopulate with recipient cells but actually become completely acellular within months of implantation. This is of particular concern as the human aortic valve is expected to be the ideal matrix for repopulation: it has the right mechanical properties and is far less antigenic than porcine matrixes. Possibly in view of these pathological findings, or possibly for other reasons, a number of investigators have attempted to repopulate these scaffolds in vitro, in advance of implantation into animal models. The CryoLife technology, however, was implanted into patients without any prior cell seeding in vitro. Indeed, technologies attractive to the valve industry are those that do not involve handling of patient cells. It is far easier to get a new valve approved by the FDA as a device, if there is no reliance on cells contributing to its long-term durability.

Bioresorbable Scaffold

This approach to valvular tissue engineering is perhaps the most conventional. The bioresorbable scaffold has been used in applications as varied as skin, bone, vessels, spinal chord, tendon, bladder, vagina, muscle, and solid organs like liver and pancreas. The concept is simple: cells of a particular phenotype are seeded on a porous material, implanted in the body, and are expected to generate the organ of interest as the scaffold degenerates. The scientific rationale for this approach is unclear and based primarily on empirical evidence that it appears to work in certain applications. Perhaps the oldest most successful application of this approach is tissue-engineered skin, dating back to the late 1970s.

This approach launched the tissue-engineering industry and remains the most, if not the only, successful product. The original scaffolds for heart valves were borrowed from skin: polylactic and polyglycolic acid and copolymers thereof. These materials have been largely abandoned for heart valve applications because they were too stiff, and newer, more compliant materials like polyhydroxyalkanoate, have been used. Natural scaffolds, like small intestine submucosa have also been used for valvular matrixes. In most cases, the valve candidates have been implanted in the pulmonary, rather than the aortic position, because the degrading scaffold cannot bear left ventricular pressures before new tissues being regenerated. Besides the option of scaffold material and shape of the leaflets, the other variable is the decision of whether to preseed in vitro or not and, specifically, how to do it. As discussed above for the acellular matrix valve, preseeding in vitro appears to yield better in vivo results.

Collagen-Based Constructs Containing Entrapped Cells

This is perhaps the least explored area and the one that involves the greatest amount of engineering. This approach is based on the relatively old observations that cells entrapped in collagen gels contract and compact the gels, increasing the density of the collagen many fold. The principle involves first mixing soluble, fibrillar collagen with the appropriate cells. After the collagen–cell mixture is neutralized, soluble collagen reassembles into fibrils and a gel is created. Cells become entrapped within the collagen gel and begin to interact with the collagen fibrils and contract the matrix, excluding water. In many ways, this in vitro contraction mimics wound healing in vivo. When the gel is mechanically constrained, the collagen fibrils align in the direction of constraint, and a highly aligned, compacted collagenous construct can thus be fabricated.

One of the early applications of this technology has been as blood vessels,where the fabrication of collagen tubes is relatively straightforward. All that is required is a rod affixed centrally within the lumen of a tube, creating an annular space that is filled with the collagen–cell mixture. Once the mixture gels, it begins to contract around the central rod and peels off the inner surface of the tube. At that point, the rod with its adherent gel coating can be removed from the tube. Because the gel coating on the rod cannot shrink in circumference, it compacts only through its thickness and along the length of the rod. Collagen fibrils thus align circumferentially, much like in the adventitia of a blood vessel. A very logical extension of this approach to heart valves is to fabricate a mold for leaflets, contiguous with the aortic annulus. This is what has been done by the laboratory of Tranquillo et al. Tranquillo developed an interesting mold with inner and outer parts that produced the shape of a bileaflet valve within a tube. Like the vessel wall, the valve leaflets also shrank along the direction of the tube, becoming shorter and developing an aligned fiber structure.

The work in the laboratory of this author is a simplification of the original work of Tranquillo. Rather than forming complex 3D structures, we have focused on generating simple, 1D strings that could be used for surgical reconstruction of mitral valve chordae or for future use in more complex constructs. We began by selecting neonatal rat aortic smooth muscle cells as the experimental cell line, because they are well known for producing considerable amounts of matrix, particularly elastin. We found empirically, however, that cells with highly varied phenotype can be extracted from the minced aortic tissues. Digesting these tissues with trypsin apparently liberates a mixture of cells, many of which are not highly contractile and thus unsuitable for use with directed collagen gel shrinkage.

The best outcomes were obtained with cells isolated by the outgrowth method, in which tissue is minced into small pieces and plated onto to dishes. The most highly motile cells apparently grow out from these tissue fragments and can be amplified over several weeks of culture. Optimal shrinkage of the constructs occurred when these cells were added to the collagen suspension at a cell-seeding concentration of 10^6 cells/mL. The collagen suspension consists of sterile acid-soluble type I collagen at an initial concentration of 2.0 mg/mL.

For our application, the cell/collagen suspension is pipetted into rectangular wells of variable geometries with microporous holders at their ends. Like the inner mandrel of the tubular constructs, the anchors at the ends of the wells prevent longitudinal contraction and allow shrinkage to occur only transversely to the long axis of the wells. This gives rise to well-aligned collagen constructs with relatively high collagen fibril density. After 8 weeks of culture, the collagen constructs have the typical nonlinear stress/strain curve of tendinous materials, an extensibility of 14%, a stiffness of 5 MPa, and failure strength of 1.1 MPa. Although the stiffness and strength are still about an order of magnitude lower than what is required, our constructs are already 10 to 100 times stronger than similar collagen-based materials fabricated previously. Ultrastructural analyses have shown that the main reason for the good strength of our constructs is the very high collagen fibril density. Because the constructs are relatively simple 1D collagen bundles, they compact from 2 directions, producing an area shrinkage ratio that is greater than 99%. In an effort to improve the strength of these constructs further, we have explored different sizes and aspect ratios, different materials for the anchors for these constructs, and different forms of application of external tension. For example, triangular-shaped holders that appear to channel the tension from the holder material to the construct lead to stronger constructs, as does the application of external forces. Dynamic loading, in particular, increases construct strength by a factor of 3. Although these constructs are clearly not ready for human use, they are nearing use in animal models. One point of concern is the use of rat collagen and rat cells, and efforts are underway to translate this technology from the rat to the sheep model, making use of sheep collagen and sheep cells. The core technology, however, is also being used in a more ambitious approach to develop a composite aortic valve cusp. This is discussed below.

Other Substrates in Early Development

The main problems with using reconstituted collagen as a substrate for tissue engineering is the observation that cells entrapped in collagen gels rapidly enter apoptosis and synthesize matrix metalloproteinases. Whereas strategies for overcoming these phenomena have been attempted (ie, mechanical loading), use of collagen alone has thus not been widely embraced. A number of investigators are exploring cell adhesion and phenotype on thin flat films of candidate materials, as a prelude to using the bulk material as a scaffold. For example, Giachelli and colleagues have explored the use of chitosan, an interesting material derived from crustacean shells and used for wound dressings, and found that mixtures of chitosan and collagen work better than either alone.

Anseth and colleagues have used crosslinked polyvinylalcohol and hyaluronan as a substrate for valve tissue engineering. Rothenburger and colleagues from Muenster, Germany, have used collagen films as scaffolds for tissue engineering and found that myofibroblasts and endothelium cocultured produced significant amounts of collagen and structural proteoglycans. Fibrin is being considered an alternative to collagen in valvular tissue engineering by a number of groups, including that of Tranquillo.

HISTORY OF EACH APPROACH: SUCCESSES AND FAILURES

Acellular Matrix Valve

Judging from the volume of literature, the acellular matrix valve has received the most interest. This is not surprising, because it seems to make the most sense from a biomechanical point of view. The aortic valve is only that: a mechanical structure designed to open and close with minimal pressure drop and reverse leakage. What is unique to the aortic valve, compared with most other connective tissues, is that it does not require any metabolic activity or self-repair to provide a good service life. Recall that human aortic valve allografts (homografts) are transplanted without any tissue-type matching and consequently become acellular within a few months. The homograft can last 20 years, essentially as a dead piece of tissue, free from any mechanical reinforcement by crosslinking agents. Indeed, it is remarkable that the homograft valves last longer than the glutaraldehyde-fixed porcine xenografts; there must be something about the microstructure and composition of the native aortic valve that makes it very resistant to mechanical fatigue. Indeed, many have studied the structural basis for this remarkable durability of the aortic valve. Although it still remains unknown exactly what features of the native valve tissue give it such remarkable durability, the importance of its internal complexity is being appreciated more and more.

Most likely, the presence of interconnected sheets of collagen, layers and tubes of elastin, highly nonlinear mechanics, anisotropy, and viscoelasticity endow the valve tissue with is unique longevity. Not knowing exactly which features are important, the logical approach is to duplicate all of them in a prosthetic device. Indeed, the use of intact porcine aortic valves has spawned the whole bioprosthetic valve industry. Early in its history, engineering analyses clearly demonstrated that the porcine xenograft, with its inherent structural complexity, is theoretically better than the pericardial valve and should thus last much longer.

Interestingly, this conventional wisdom ultimately proved to be quite wrong: a particularly good design of the pericardial valve (the Carpentier-Edwards valve) is remarkably long lived and appears to be more durable than glutaraldehyde-fixed porcine valves. Now it could be argued, of course, that a glutaraldehyde-fixed aortic valve, mounted on a stent, is hobbled somehow and its theoretical advantage over the pericardial valve disappears. That could very well be the case, because some studies have shown certain disadvantages of glutaraldehyde-fixed aortic valve cusps relative to similarly prepared bovine pericardium. The intact aortic valve may thus still have the upper hand when finally revitalized with freshly seeded cells. Regrettably, there have been few, if any successes, with this approach. Rather than happily repopulating the waiting matrix, the cells that come into contact with acellular valves have reacted badly and destroyed the intricate microstructure and its associated perfect mechanics. The reality of this, unfortunately, has come to light only recently. For almost 20 years, research groups around the world have independently tried various chemical processing approaches and published the successes in short-term animal models, but none led to any real clinical success.

The first reports that these protocols ultimately fail in long-term animal studies were only anecdotal at first, shared among scientists during personal discussion at the various heart valve conferences. No negative results were actually published until Hilbert, a well-recognized valve scientist working at the laboratories of the NIH, did the "ultimate" comparative study of various extraction protocols in a long-term sheep implantation study. Hilbert copied the protocols of a number of investigators and implanted 2 sets of differently processed valves into the sheep model, as pulmonary artery interposition grafts. After 20 weeks of implantation, the valves were explanted and examined grossly and histopathologically. With some variation between treatment protocols, all valves underwent considerable tissue overgrowth and infiltration with inflammatory cells. There was also some evidence of aneurismal dilatation.

The second such report came from the laboratory of Stock, who reported similar complications with his valves, even though these valves were seeded with cells before implantation and were thus more "ready" for implantation than the valves prepared by Hilbert. It could be argued, however, that the sheep model was destined for failure and that the acellular xenograft approach would work in the human condition. The reason for this argument is that it has become clear over the past decade that the sheep model generates an exuberant fibrotic response to valve implants. Valves that are implanted in sheep overgrow rapidly with fibrotic tissue, certainly much more than they do in humans.

Indeed, a painful lesson learned in this area was experienced by Sulzer Carbomedics, whose Photofix-˜ pericardial valve developed severe abrasions to the leaflets against the sewing cuff during its clinical trials in the mid-1990s. The reason for the leaflet abrasion was attributable to a very specific design flaw, which was not anticipated. Indeed, this flaw did not manifest in the preclinical sheep studies because the sheep rapidly covered the sewing cuff with pannus and thus protected the delicate valve leaflets from the more abrasive sewing cuff material. It could be thus argued that testing acellular matrix valves in the sheep model is a waste of time and destined to fail, because of this severe fibrotic response.

Perhaps this was the motivation behind the use of the CryoLife Synergraft in patients directly, without any reported long-term animal studies, something quite unusual for the heart valve field. The CryoLife approach was similar to the others: the removal of cellular antigens through dissolution and extraction. Unlike the somewhat harsh chemistry of the original Brendel and Duhamer approach, and most of those who tried to improve on it, CryoLife favored a more gentle approach that has not been revealed publicly but most likely involves multiple freeze-thaw cycles, cell lysis through variation in osmotic pressure, use of enzymes such as DNase, and prolonged extraction in aqueous solutions.

The CryoLife product was thus expected to be the least affected and the most mechanically sound acellular matrix. In October of 2000, CryoLife Inc obtained CE Mark approval to sell the Synergraft product in Europe to a limited group of patients who had few alternatives: the neonate with congenital valvular malformations. Although the numbers are not widely reported, a number of patients in Vienna, Austria, received these valves.

Within a few weeks to months, many of these children began to experience serious valvular complications. Although some remained apparently unaffected, many valves became highly regurgitant, and some children died. CryoLife rapidly withdrew the product from the market and disclosed the negative clinical outcomes at the 2004 Florence Heart Valve meeting. In their 2003 article, Simon et al reported that 1 patient (aged 7 years) died 7 days after implantation of the valve, 1 patient (aged 9 years) died 6 weeks after surgery, and 1 patient (age 2.5 years) died 1 year after surgery. Death resulted from various cardiac complications related to inflammation, valve rupture, and stenosis. One patient was reoperated on 2 days after the primary surgery, in view of what happened to the others. All valves showed severe inflammation, both inside and out, fibrosis, encapsulation, perforation, and deterioration of the leaflet tissues. Grossly, this reaction was not markedly different from that shown by Stock and colleagues in the sheep model. The sheep model was thus not to blame; it was the matrix itself after all. Apparently, there is some abnormal signaling generated by the acellular matrix that induces the invading cells to remodel, fibrose, and contract what otherwise appears to be a perfect, mechanically sound valve matrix. The exact reasons for this remain unclear.

Although its first clinical use was disastrous, proponents of this approach remain undeterred. Apparently, another variant of this approach is doing well in Berlin, Germany, and is being implanted by Dohmen, Konertz, and colleagues with 50 implants in patients for more than 2 years. In this series, 1 patient died postoperatively, 2 required reoperation for valve-related complications, and the rest of the patients appear to be doing fine. The published short-term results in adults (17 to 70 years; mean, 46 years) therefore appear to be good, and the surgeons are satisfied with this technology. Personal reports obtained at the Society for Heart Valve Disease Meeting in Vancouver in June of this year, however, indicate that a number of their patients have indeed died from valve-related complications. In their article, the authors also disclose that Konertz and Dohmen are shareholders of AutoTissue GmbH, the company that produces these valves.

Resorbable Scaffold

The failure of this approach is much less dramatic, as apparently none of these scaffolds have been tried in patients. Interestingly, the main and still dominant reason for favoring the bioresorbable polylactic/polyglycolic acid (PLA/PGA) scaffold approach is the FDA approval of the materials for implantation in the US, as bioresorbable sutures. Although this may be true, it most likely does not lessen the regulatory burden of a device constructed from these materials. Valves are life-sustaining class III devices and must be taken through the full set of preclinical and clinical trials before market approval. Perhaps it is thought that as limited-use surgical materials, valves fabricated from these matrixes could be used in limited volumes by surgeons directly, as part of an investigative study in their hospital. Perhaps the attractiveness of this approach is the legacy of Robert Langer who pioneered the resorbable matrix approach in the artificial skin product and launched the tissue engineering industry. Perhaps the entire approach is flawed. Jeff Hubbel, a noted chemical engineer with expertise in developing materials with controlled release chemistry noted recently at a conference that biological matrixes do not degrade through hydrolysis, but rather through proteolysis. Accordingly, he has been engineering his matrixes to be cleaved by enzymes. Hydrolysable matrixes therefore are clearly not biomimetic and their use as biodegradable substrates should thus give rise for concern. Matrixes cleavable by proteolytic enzymes, however, are also far from clinical reality and may indeed experience the same fate as the others.

The fate and the failure mechanism of the bioresorbable matrix approach to heart valves, unfortunately, is less clear than it is for the acellular xenograft. Like for the acellular matrix valves, there have been no clear reports of "failed experiments." Personal communications with the principal investigators, however, confirm that fibrosis, retraction, and incompetence have hampered the progress of valves based on resorbable matrixes. No good histological pictures of failed valves appear in the literature, and thus little can be learned from the 10 years of experience with this approach.

Each group working in this field appears to publish promising short-term results with a particular matrix material and yet abandons it in favor of a more promising one with no report of how the old material fared. What remains unclear to most of the scientific community is specifically how the previous approach failed. Although PLA/PGA was abandoned by the pioneers in this field—the group of Vacanti and Mayer of Harvard/Children's Hospital—in favor of a polyhydroxyalkanoate, a better, more compliant material, other groups around the world continue to work with PLA, most likely because it is the only FDA-approved, clinically popular material. As the various research groups working in this field continue to use resorbable materials, the questions that remain can be unpleasant: Are PLA/PGA and their variants inappropriate materials, or have they simply been used improperly by the first groups?

Are the other groups going to see the same results and waste a lot of time? What is the responsibility of the research community in helping others avoid approaches that failed and perhaps choose a better path? In an environment where the prize is a piece of a \$1 billion annual market, competition for attention and credibility is clearly felt at scientific meetings. Reporting about failures, rather than successes, also does not make for exciting reading and negative results seldom get published in mainstream journals. The ability for scientists to learn from the mistakes of the past is therefore quite limited.

HYBRID APPROACHES

A possible alternative to the acellular valve and the bioresorbable matrix approaches is the fabrication of complex structures by manipulating biological molecules. With sufficient fidelity, one could potentially fabricate structures as complex as the aortic valve cusps. This approach is a derivative of that pioneered by Tranquillo, except that the directed collagen shrinkage method is simplified to a process that forms essentially 1D strings of collagen. These collagen fiber bundles can then be used as building blocks for the development of a composite aortic valve cusp. Whereas collagen structures in heart valves can be found in the form of sheets, most of the load-bearing components of the valve cusp are relatively thick, dense collagen fiber bundles. Having developed this technology already the next step was to develop the other components required to make this approach work.

The aortic valve consists of the three basic building blocks of all connective tissues: collagen fibers, elastin sheets, and glycosaminoglycan matrix schematically arranged. The collagen exists to bear tensile load and provide some ultimate stiffness and strength to the valve, so that it can withstand diastolic loads. The elastin matrix exists to return the collagen structures back to their resting states between loading cycles, and the glycosaminoglycans likely maintain hydration and the intrinsic viscoelasticity of the tissue. For the latter component, hyaluronan was chosen as the working material.

Hyaluronan is a glycosaminoglycan polymer with a repeating disaccharide structure (glucuronic acid–b1,3-*N*-acetylgalactosamine-b1,4-)*n*, where *n* can reach 25 000 or more. In solution, hyaluronan forms large, random coil structures that occupy large solvent volumes. When constrained within a matrix, like a collagen network, hyaluronan exerts a swelling pressure that traps water, gives tissues compressive resistance, and imparts viscoelastic properties. Hyaluronan also plays a role during embryonic cardiac development as invading cells transform "cardiac jelly" into specialized cardiac structures, like the myocardium and the cardiac valves. Hyaluronan is exceptionally biocompatible.Unlike collagen that has both tissue- and species-specific markers, hyaluronan exhibits structural homology across species. For example, hyaluronan made by bacteria is the same as hyaluronan made by humans. Because of its interesting viscoelastic properties and its broad biocompatibility, hyaluronan has been used in a number of clinical applications and also as a scaffold for heart valve tissue engineering by others.

The formulation of hyaluronan gels used in our laboratory is based on a modification of a previously patented protocol. Details of the preparation are published elsewhere, but in brief, the process involves mixing commercially available sodium salt of long chain hyaluronan with 1 mol/L NaOH at low temperatures and crosslinking with divinyl-sulfone. These gels can have a stiffness as high as 30 kPa, similar to that of the elastin structures of the aortic valve cusps, and thus can form highly hydrated, elastic sheets. It is expected that appropriately cast sheets of crosslinked hyaluronan can find use in the central spongiosa layer of the aortic valve cusp. Two spurious findings occurred when working with the collagen constructs and the crosslinked hyaluronan; both stimulated the formation of elastin sheets. Elastin sheets formed spontaneously around the periphery of the collagen constructs and on hyaluronan substrates texturized through dehydration and texturization with ultraviolet light.

Whereas the growth of elastin sheets on the collagen constructs is straightforward, growth of elastin sheets atop the hyaluronan gels has been difficult to reproduce. The success of this technique appears to be highly dependant on the specific cell type, gel formulation, and degree of surface texturization by UV light irradiation. Standardization of these protocols is an ongoing process in our laboratory, and no animal implants have yet been done.

STEM CELLS AND OTHER FUTURE PROSPECTS

The cell is clearly an important component of the tissue-engineered heart valve. A very good review of the types of cells used in tissue-engineered valves is provided by Flanagan and Pandit. Much of the work identifying the phenotype of interstitial cells in the aortic valve has been done outside the US, in the laboratories of Yacoub, Gerosa, and Boughner. Which cell to use for seeding scaffolds remains unclear. Stem cells and various other progenitor cells are being increasingly used in tissue-engineered valve applications. The 2 main cell types are mesenchymal stem cells and circulating endothelial progenitor cells. These cells are harvested from either experimental animals or patients, expanded in culture, and then seeded on the various valve leaflets substrates, be they the acellular matrix valves or the resorbable scaffolds. From presentations at recent conferences, all of these studies are very preliminary, showing no real differences in histological morphology over conventional interstitial cells used in previous studies.

Many investigators believe that cells are "smart," somehow recognizing the substrate and behaving in the appropriate way. The fibrotic overgrowth and failure of valvular matrixes described above clearly points to the contrary. With the emergence of stem cell science, many investigators believe that stem cells are going to be the "silver bullet" and are going to be smarter and act appropriately on the substrates. Embryonic stem cells, when injected into infracted hearts, have not behaved "smartly," did not rebuild the damaged myocardium, and, instead, created calcific deposits or teratomas. There is no evidence that undifferentiated stem cells will behave in a more intelligent way on valvular substrates. Most likely, considerable effort will need to be expended to differentiate embryonic stem cells along a valvular lineage in advance of seeding on valvular substrates, before the promise of stem cells can be realized in this field. Stem cell, however, are clearly the wave of the future. Considerable evidence exists in the literature that matrixes implanted without cells resorb, fibrose, and fail to produce any clinical benefit, whereas matrixes seeded with relevant cells incorporate and offer therapeutic benefit. The work of Atala in the field of urology is perhaps the best example of cell-seeded matrixes providing real clinical benefit to patients in augmenting large soft tissue defects. Stem cells, once differentiated to the proper end point, are expected to provide a broader source of autologous cell lines, along with the appropriate matrixes, for therapeutic use in the cardiovascular field.

The use of autologous cells, cell expansion, preimplantation culture, and other cell "management," however, is not the preferred approach in the medical device and therapeutics industry. Because of regulatory and cost issues, a tissue-engineered valve that does not require any cell management is the preferred choice for medical device companies. This is the reason for which CryoLife and originally St Jude and Advanced Tissue Sciences and currently Medtronic have pursued the acellular matrix approach; it can be qualified, manufactured, and sold as a device, a far simpler approach than the biological or combination product that most other tissue engineering therapies are likely to follow.

The fact that the CryoLife approach failed, however, has not deterred others from trying it. The literature clearly shows that the majority of research projects prefer the acellular matrix approach, particularly now with the greater promise of stem cells.

8

Microscale Technologies for Tissue Engineering and Biology

Each year in the U.S., millions of people suffer from a variety of diseases that could be aided from therapies such as organ transplantation. However, despite the widespread need for transplantable tissues, many patients die while waiting for donor organs. It is from this need that the field of tissue engineering has emerged. Tissue engineering is an interdisciplinary field that involves the use of biological sciences and engineering to develop tissues that restore, maintain, or enhance tissue function. Tissue engineering has particular advantages over other therapies such as drugs because it can provide a permanent solution to the problem of organ failure. In general, there are three main approaches to tissue engineering: (*i*) to use isolated cells or cell substitutes as cellular replacement parts, (*ii*) to use acellular biomaterials capable of inducing tissue regeneration, and (*iii*) to use a combination of cells and materials (typically in the form of scaffolds).

Despite significant advances in tissue engineering, which have resulted in successful engineering of organs such as skin and cartilage, there are a number of challenges that remain in making off-the-shelf tissue-engineered organs. These barriers include the lack of a renewable source of functional cells that are immunologically compatible with the patient; the lack of biomaterials with desired mechanical, chemical, and biological properties; and the inability to generate large, vascularized tissues that can easily integrate into the host's circulatory system with the architectural complexity of native tissues.

Microscale technologies are potentially powerful tools for addressing some of the challenges in tissue engineering. MEMS (microelectromechanical systems), which are an extension of the semiconductor and microelectronics industries, can be used to control features at length scales from <1 μm to >1 cm. These techniques are compatible with cells and are now being integrated with biomaterials to facilitate fabrication of cell◎ material composites that can be used for tissue engineering. In addition, microscale technologies allow for an unprecedented ability to control the cellular microenvironment in culture and miniaturize assays for high-throughput applications. In the past few years, microfabrication has been increasingly used in biomedical and biological applications, partly because of the emergence of techniques such as soft lithography to fabricate microscale devices without the use of expensive "clean rooms" and photolithographic equipment. Soft lithography is a set of microfabrication techniques that use elastomeric stamps fabricated from patterned silicon wafers to print or mold materials at resolutions as low as several hundred nanometers. Therefore, soft lithography can be used to control the topography and spatial distribution of molecules on a surface, as well as the subsequent deposition of cells. Soft lithographic methods can also be used to fabricate microfluidic channels and scaffolds for tissue engineering in a convenient, rapid, and inexpensive manner. In addition, photolithography, a technique in which microscale features are fabricated based on selective exposure of a material to light, can also be used for microfabrication of tissue engineering structures.

In this review, we will discuss the use of microfabrication technologies as they relate to tissue engineering in addressing challenges such as vascularizing engineered tissues, fabricating 3D structures, directing stem cell fate, and controlling cellular microenvironment. In addition, we will discuss the use of these technologies for providing significant insights into cellular behavior *in vitro* and for creating physiological microenvironments in culture. We will first discuss the use of microscale approaches as they apply directly to tissue engineering in fabricating scaffolds and bioreactors. We will then discuss the use of these technologies as they apply to tissue engineering indirectly through their use in controlling cellular microenvironment.

MANUFACTURING MICROSCALE MEDICAL DEVICES FOR FASTER TISSUE ENGINEERING

Now a team of researchers from the Laser Zentrum Hannover (LZH) eV Institute in Hannover, Germany, and the Joint Department of Biomedical Engineering at the University of North Carolina at Chapel Hill and North Carolina State University have modified a manufacturing technique called two-photon polymerization (2PP) to create finely detailed structures such as tissue scaffolds more quickly and efficiently than was previously possible. The new technique, which the team describes in a paper published this week in the Optical Society's (OSA) open-access journal *Biomedical Optics Express*, could help pave the way to more wide-spread clinical use of microscale medical devices.

Many important biological functions take place on the microscopic level and as medical research advances into this Lilliputian realm, scientists have turned to precise techniques such as 2PP to create the tiny tools necessary to manipulate cells and other miniscule structures. In current-generation 2PP technology, a laser pulse that lasts approximately one quadrillionth of a second sends a burst of energy into unset resin, causing the molecules around the pulse to fuse together into two adjoining cone shapes. By focusing on multiple points in succession, 2PP can build up complex 3D structures, cone-shaped block by cone-shaped block.

2PP can be used to manufacture devices from a wide range of base materials and does not require extreme temperatures, harsh chemicals, or cleanroom facilities, but its main drawback is long fabrication times. Like in a tiled mosaic, small 2PP building blocks can create a richly detailed design, but if you want a large structure, like a tissue scaffold that could mimic natural body parts, it can take a long time to lay all the pieces together.

"Blood vessel networks can be several centimeters in length, but walls of the smallest branches (capillaries) are only a few micrometers thick. The same applies with any tissue. Many tissues may be large, but they all have important features on the microscale," says team member Shaun Gittard of the LZH. The team notes that using conventional 2PP to manufacture the tissue scaffolds for such structures could be prohibitively slow. They address the problem by using a computer-controlled hologram to split the 2PP laser into multiple beams, creating up to 16 different focus points that can work simultaneously.

"As an example, take the time for fabricating a single layered, 1-millimeter square with 100 nanometer resolution," the authors write. "With conventional single-focus 2PP at one millimeter per second, the fabrication time would be 2 hours and 47 minutes. In contrast, with 16 foci this same area could be scanned in merely 10 minutes." Or, in other words, many foci make light work.

The team first tested their multiple foci system by creating 16 miniature Venus statues, each so small as to be invisible to the human eye. "The Venus is kind of a logo of our research group," says Gittard. "We have used it as a familiar demonstration structure for various fabrication techniques." In addition to replicas of classic Greek artwork, the team also used the new technique to manufacture cylindrical tissue scaffolds and an array of microneedles. Less than a half millimeter wide, rocket-shaped microneedles can be used to provide painless injections or take blood samples, notes Gittard. "One of the biggest promises in the future is real-time, pain-free glucose sensing and insulin delivery for treating diabetes," he says.

For now the team has only used the multiple beams to create multiple copies of the same structure. Their next goal is to use the system to produce one large, complex 3-D structure, which is a more complicated task since it requires moving the relative placement of the different foci during the fabrication process, Gittard says. "The ability to produce large-scale devices with sub-micron features is exciting, as many cell features, such as organelles, are on this size scale," the authors write. Gittard explains that such detailed features could be used to control cell attachment and alignment, which is important since cell orientation affects function in a number of tissues, such as blood vessels, nerves, bone, and muscle.

APPLICATION OF MICROTECHNOLOGIES FOR THE VASCULARIZATION OF ENGINEERED TISSUES

Progress in the development of large tissue-engineered organs has so far been limited by the inability to generate sophisticated three dimensional (3D) structures comprised of a functional vasculature. The vascular system is a dynamic environment comprised of a variety of cell types that constantly remodels itself under the influence of endothelial, immune, nervous and endocrine cells . Vascular growth and remodeling are coupled with developmental and wound healing processes as well as the progression of various pathologies such as inflammation, cardiovascular diseases and cancer. Most of these processes depend on endothelial cells, which line the interior of blood vessels and form the endothelium.

This interface between circulating blood and the surrounding tissues is responsible for proper solute transport and molecular exchange. It ensures the delivery of sufficient oxygen and nutrients to cells to maintain tissue homeostasis. Cells *in vivo*are found to be at most a few hundred microns away from the nearest capillary or blood vessel. Beyond this distance, diffusion is not effective and tends to reduce cell survival and function. Therefore, the inability to adequately vascularize engineered tissues results in inefficient transport of nutrients and metabolites and often leads to cell death and tissue necrosis. Moreover, vascularization of engineered tissues plays an important role in graft perfusion and is also crucial in facilitating the integration of the implanted material with the host vasculature .

The fabrication of tissue constructs often requires cell seeding of 3D scaffolds. These scaffolds are generally made of gels, foams or fibrous meshes and usually have basic macroscale properties that enable cell adhesion, migration and proliferation . Although these properties are often sufficient to allow the formation of functional connective tissues such as skin, bladder and cornea and 3D tubular structures like blood vessels and urinary tract, most tissue engineering approaches still lack the capability to sustain the growth of thick engineered organs . Skin and cartilage have been among the first engineered tissues ready for clinical applications since they do not require extensive internal vasculature to survive *in vivo*. However, the fabrication of complex organs such as the heart or liver requires adequate vascular supply to ensure the survival of specialized cells within their 3D structure. To achieve this level of functionality, it is necessary to integrate a network of vessels ranging from a few microns to several millimeters within engineered tissues. The incorporation of this microcirculation into tissues and organs represents a considerable challenge which includes the engineering of vascular conduits having micron scale dimensions. It also requires a functional endothelium to improve vascular activity and avoid thrombosis, as well as other specialized cell types performing the physiological tissue function of interest. Various approaches have been proposed to design scaffolds comprised of a vascular network analogous to capillaries. These approaches rely on the release of growth factors from the scaffolding material, the seeding of endothelial cells in the scaffold to promote angiogenesis or the use of microfabrication technologies to engineer branched microfluidic channels inside biocompatible materials . Regardless of the application, these technologies all aim at improving mass and fluid transport as well as oxygen diffusion in engineered tissues produced *in vitro*.

MICROVASCULARIZATION OF ENGINEERED TISSUES THROUGH ANGIOGENESIS AND INOSCULATION

The need for adequate solute transport in cell-seeded scaffolds is essential for tissue survival and function. A key approach in attempting to induce the growth of a vascular network within 3D engineered tissue has been the incorporation of growth factors into the scaffolding material. It was shown that a macroporous scaffold, obtained by particle leaching, freeze drying or other pore forming technologies and functionalized with growth factors such as vascular endothelial growth factor (VEGF), basic fibroblast growth factor (bFGF) and platelet-derived growth factor (PDGF), can trigger the formation of vascular structures following *in vivo* implantation .

Both natural and synthetic scaffolding materials have been loaded with these pro-angiogenic molecules, leading to the sprouting of capillary beds within the constructs. However, the lack of directed growth of blood vessels to enable interconnectivity between the capillary networks still remains the biggest challenge of this technology.

Microvascular structures incorporated in tissue engineering scaffolds prior to their implantation can also be obtained by the co-culture of endothelial cells with the cell types of interest regarding tissue function. This cell-based approach uses the ability of endothelial cells to release growth factors and promotes the formation of capillaries *in vitro*. The resulting angiogenesis phenomenon can be explained by the remodeling that occurs within the construct, which is driven by the endothelial network that activates the release of pro-angiogenic factors in the construct. The remodeling of the extracellular matrix (ECM) allowing the formation of vasculature is driven by matrix-metalloproteases (MMPs), regulating cell proliferation and migration within the tissue . Using a skin model, it has been originally demonstrated that a capillary network can be successfully produced by culturing human umbilical vein endothelial cells (HUVECs) into the dermal part of the engineered substitute leading to the formation of long-lasting and perfused blood vessel networks following *in vivo* transplantation . This inosculation enabled the perfusion of the tissue engineered capillary network by the vasculature of the host following implantation. The cell-based approaches have led to many interesting results including vascularized muscle, cardiac tissue, bone and blood vessel, all of which have been implanted in animal models and have shown improved perfusion and inosculation with the host vasculature. Similar results were obtained when HUVECs were cultured in a 3D collagen ECM *in vivo*, where a stable tree-like structure of a branched network was observed for an extended period of time . Several other recent studies also suggested that mesenchymal stem cells and endothelial cells have the ability to interact together to form a stable vascular network of capillaries both under *in vitro* and *in vivo* conditions .

In vivo studies also demonstrated that the presence of vascular structures formed *in vitro* greatly accelerated inosculation of the implanted tissue with the host vasculature. Results from L'Heureux*et al.* have shown that the formation of a vasa vasorum in the wall of a tissue engineered blood vessel (TEBV) occurred 3 months following implantation *in vivo* . Similarly, Guillemette *et al.* showed that a TEBV comprised of a vasa vasorum engineered *in vitro*(TEBVwVV) allowed for complete tissue integration and functional vasa vasorum activity after only 2 weeks *in vivo* . These studies have provided evidence for the need to incorporate a capillary network into engineered tissues prior to implantation, showing improved and accelerated tissue integration and capillary-like structure formation which enabled connections with the host vasculature *in vivo*.

Therefore, rapid formation of a vascular network between the engineered tissue and the host at the transplanted site, similar to the process observed during wound healing, represents a promising solution to provide implanted cells with adequate supply of oxygen and nutrients. However, this strategy does not provide a sustainable solution that can enable both perfusion *in vitro* and inosculation *in vivo*. In addition, these approaches did not result in the creation of organ scale constructs *in vitro*. In an attempt to engineer vasculature into engineered tissues and organs, microscale technologies and microfluidics systems have emerged as efficient tools to create easily perfusable channels in biocompatible and biodegradable 3D scaffolding materials.

THE USE OF MICROTECHNOLOGIES TO ENGINEER VASCULARIZATION IN VITRO

The application of microtechnologies to biomaterials can be used to reproduce the capillary network and allow the flow of culture medium through a construct during *in vitro* studies. Unlike angiogenesis and inosculation approaches, microfabricated devices with integrated microvasculature can be optimized to provide a uniform distribution of flow and mass transfer across the scaffolding material and thus provide the cells with an adequate supply of nutrients.

Soft lithographic and micromolding processes have been used to create microfluidic devices consisting of branched networks that can be connected to perfusion systems *in vitro* . These technologies have been applied to polymers such as poly(dimethyl siloxane) (PDMS), poly-lactic (co-glycolic acid) (PLGA) and polyglycerol sebacate (PGS) in which channel networks can be perfused and seeded with vascular cells . These techniques have been shown to be useful to regulate the formation of vascular networks in a precise and efficient manner. The design of functional microvascular networks involves the integration of multiple parameters such as the geometry of branching and bifurcations, fluid mechanics, mass transport and structural rigidity.

These properties are of utmost importance since they greatly influence the stability, oxygen and nutrient distribution and therefore the functionality of engineered tissues. Microfluidic systems can effectively transport solutes in capillary channels ranging from a few millimeters down to micrometers . The control of fluid mechanics and mass transport over this wide range of dimensions has been used to study bioactive molecules and therapeutics in cardiovascular research and tissue engineering . Sophisticated devices have recently been fabricated to reproduce a lung assist apparatus allowing the blood being perfused within the microchannels to be oxygenated by flowing through many parallel capillary-like channels analogous to the native lung architecture .

Although optimization of the gas transfer membrane and characterization of the blood flow in the device are still needed, this is a good example demonstrating the potential of microfabrication technologies to generate vascularized platforms for tissue engineering. Similarly, it was shown that microvascular cells could be seeded in the device to form a confluent endothelium on the walls of the vascular channels . These studies demonstrated that microfabricated devices comprised of a fluidic network modeled on human vasculature can be successfully inosculated *in vivo* . Even though previous work has shown that microscale channels can be engineered *in vitro,* there is actually no available method to consecutively branch multi-dimensional channels inside a scaffold . Top-down fabrication processes are inherently planar in nature and therefore 3D structures mostly result from stacked 2D structures which are comprised of channels having rectangular cross sections .

Moreover, most microfabrication techniques have been developed for materials that are unable to sustain cell encapsulation, which represents a limitation of this approach to generate large vascularized tissues . Few attempts were made to engineer perfused microfluidic devices in which two different cell types can be cultured, but a vascularized tissue with parenchymal cells has yet to be created and there are still no methods to build sustainable large cell-laden structures with multi-dimensional branched networks.

Other microscale fabrication approaches such as bioprinting and stereolithography are currently being investigated to create 3D branching vascular networks . However, these methods not only require specialized facilities and expensive equipment, but the fabrication processes involved are usually time consuming.

Biomimetic Approaches for Engineering Tissue Vascularization

Cells in the body are in contact with a complex 3D environment comprised of a combination of soluble factors as well as the ECM and basement membrane proteins found in the tissue in which they reside. Most tissues consist of multiple cell types organized into hierarchical structures that allow them to regulate their function. Modular tissue engineering, or bottom-up approaches, have recently emerged as powerful fabrication methods to generate 3D structures that mimic this organization and that recapitulate tissue structure and spatial resolution . These techniques aim to generate biomimetic mesoscale structures by engineering microscale components and by using them as building blocks to fabricate larger tissue structures . This approach has been used to control the cell microenvironment and the macroscale properties of relatively large and complex engineered tissues .

Microgels are microscale hydrogels fabricated by merging microscale fabrication and hydrogel chemistry. They exhibit properties similar to native ECM, can sustain cell encapsulation and have tunable geometrical, mechanical and biological properties which make them excellent candidates for tissue engineering applications. Based on these characteristics, cell-laden microgels were fabricated and then assembled into 3D tissue constructs to create precise microarchitecture containing repeated functional units that mimic the *in vivo* structure of tissues and organs . The directed assembly of microgels can be driven by simple thermodynamic processes in a two-phase oil-aqueous reactor . Recent work has also shown the potential of modular assembly to produce vascularized tissues *in vitro.* Arrays of microgels with precisely defined structures and channels have been produced by photolithography and were assembled in a controlled manner resulting in 3D structures with multidimensional interconnected lumens resembling native vasculature. The validation of this biofabrication process was assessed by placing endothelial and smooth muscle cells inside the microgels following a precise and concentric fashion mimicking the organization of blood vessels. The complete assembly was later strengthened by a secondary crosslinking step and it was shown that these assemblies could be perfused with fluids .

One of the challenges of directed self-assembly technology resides in the difficulty to scale-up the tissue produced. Larger structures have recently been built using lock-and-key principles and modular assembly . Ongoing work is currently focusing on long-term perfusion aiming at developing mature and functional vasculature *in vitro*. Maintaining physiological functionality and blood flow in a high-density vascular network with optimized oxygen transfer conditions is critical to keep the tissue in an appropriate state of homeostasis. This biomimetic approach can also be utilized in the design of organ-on-a-chip technologies aiming at the fabrication of precise and reliable small scale models that can later be used for drug screening and physiological *in vitro*studies .

MICROSCALE APPROACHES FOR TISSUE ENGINEERING

Cell-Seeded, Microfabricated Scaffolds

In many tissue engineering applications, scaffolds are used to provide cells with a suitable growth environment, optimal oxygen levels, and effective nutrient transport as well as mechanical integrity. Scaffolds aim to provide 3D environments to bring cells in close proximity so that they can assemble to form tissues. Ideally, as the scaffold is degraded, the cells deposit their own extracellular matrix (ECM) molecules and eventually form 3D structures that closely mimic the native tissue architecture. Currently, tissue engineering scaffolds are prepared by using a variety of techniques, such as solvent casting and particulate leaching. However, scaffold properties such as pore geometry, size, interconnectivity, and spatial distribution depend on the fabrication process rather than design. The inability to generate desired scaffolds has hindered the construction of engineered tissues that are larger than a few hundred micrometers due to oxygen diffusion limitations.

Microfabrication approaches have been used to engineer the desired microvasculature directly into the tissue engineering scaffolds. Initial experiments used micromachining technologies on silicon surfaces to generate vascularized systems. Subsequent work on the replica molding of biocompatible polymers from patterned silicon wafers has resulted in the fabrication of biocompatible scaffolds. Originally, such scaffolds were fabricated from poly(dimethylsiloxane) (PDMS), a transparent, biocompatible elastomer that is widely used in replica molding. More recently, microfabricated capillary networks have been fabricated out of biodegradable elastomers such as poly(DL-lactide-*co*-glycolide) (PLGA) and poly(glycerol sebacate) (PGS). PLGA has been used to generate biodegradable microfluidic systems and microfabricated scaffolds by superpositioning and stacking multiple layers of fabricated structures. These artificial capillary networks were coated with fibronectin and seeded with endothelial cells, which grew to confluence within a few days. However, there are potential disadvantages with the use of PLGA, including its rigid mechanical properties and bulk degradation kinetics. Thus, to improve the mechanical properties and to allow for surface erosion instead of bulk erosion of biodegradable scaffolds, other polymers such as PGS have been used. PGS films were patterned with the desired microvasculature and bonded to a flat film to create a capillary network.

Layer-by-layer microfluidic patterning has also been used to generate scaffolds. In this method, cells and matrix biopolymers were flowed through channels with controlled flow rates. By sequential deposition of cells and matrix on particular regions within the microchannels, 3D structures were generated with controlled location of the deposited cells. Alternative methods of fabricating scaffolds with microscale and nanoscale resolution include 3D printing, microsyringe deposition, tissue spin casting, and electrospinning of nanofibers. In 3D printing, polymer particles and salt are printed by using a bonding agent, which forms a porous scaffold once the salt is dissolved. More recently, rapid prototyping methods such as soft lithography and microsyringe deposition have been used to fabricate PLGA and polyurethane scaffolds composed of multilayer structures with controlled resolution. Although a comprehensive review of these technologies is beyond the scope of this report, they are emerging as powerful methods of fabricating tissue engineering scaffolds.

Spatially Regulated Hydrogels and Scaffolds.

Many biological processes are regulated by spatially dependent signals. For example, gradients of molecules are commonly used in the body to regulate cell migration, axon extension, angiogenesis, and differentiation. Therefore, controlling the spatial location of molecules on a surface or throughout a material could be potentially beneficial for tissue engineering. A common approach to generate gradients is to release molecules from a source to form a concentration gradient over time as the molecule diffuses away from the source. However, these gradients are unstable, and it is difficult to control their shapes.

Therefore, methods for conjugating biomolecules to materials have been used to increase the stability of signaling molecules in a spatially controlled manner. For example, by using lasers, specific regions within an agarose gel were tethered with RGD (Arg–Gly–Asp) peptide, which allowed for neurite extension within peptide-modified regions. By using similar techniques, multicomponent, spatially patterned, photocrosslinkable hydrogels were fabricated to localize growth factors within hydrogels.

Microfabrication approaches such as microfluidics provide an attractive alternative to these technologies because of their availability and cost effectiveness. Recently, the ability to pattern fluids within microchannels has been merged with photopolymerization chemistry to form spatially oriented hydrogels. As a result, gradients of the photocrosslinkable monomers were formed within microfluidic channels and subsequently gelled by exposure to UV light. Hydrogels were synthesized with gradients of signaling or adhesive molecules or with varying crosslinking density across the material with the ability to direct cell behavior such as migration, adhesion, and differentiation. In addition to bulk characteristics of scaffold materials, surface properties also have a significant effect on cellular behavior such as cell adhesion and migration.

Microcontact printing has been used to modify surfaces of scaffold materials such as PLGA, poly(lactic acid), and chitosan with variations of polyethylene glycol (PEG) or other biological ligands, such as RGD or biotin. In addition, other techniques such as microfluidic patterning have been used to generate patterns on various tissue engineering surfaces such as PLGA. In most cases, these techniques have been limited to 2D substrates, and their incorporation to porous scaffolds remains a challenge. Microfabrication approaches have also been used to control substrate topography with resolutions as small as 50 nm to control various cell fate decisions. For example, it has been seen that surface topography of tissue culture substrates can significantly affect cardiomyocyte orientation. In these cultures, topographical features that were a few microns across were used to orient cells and enhance their function.

Furthermore, because microtextured and nanotextured substrates have been shown to significantly influence cell adhesion, gene expression, and migration, such features can be incorporated into microfabricated tissue engineering scaffolds to provide functional cues to cells. For example, topographically patterned PLGA surfaces have been shown to induce alignment and elongation of smooth muscle cells and to enhance the adhesion of several cell types such as endothelial cells and smooth muscle cells. Although these approaches have not been easily incorporated into 3D PLGA scaffolds, it may be possible to stack PLGA layers with the desired topographical features to induce specific cell responses.

Controlled Microbioreactors

Bioreactors could be useful in tissue engineering either for growing cells and tissues or as extracorporeal devices for liver and kidney diseases. Microfluidic bioreactors are potentially advantageous for cellular applications because they provide a large surface-area-to-volume ratio, as well as many other biomimetic properties. Recently, microreactors have shown promise in applications where conventional bioreactors have failed to provide the cells with adequate nutrients and oxygen.

For example, multilayer PDMS microfluidic networks have been used to culture hepatocytes *in vitro* . Inside these reactors, hepatocytes were maintained for many days as they spread and grew to confluency within the channels. In addition, at least 10 layers have been stacked, indicating that such technology is scalable. In another approach, the combination of 3D architecture and fluid perfusion has been used to mimic liver sinusoids. Silicon microfluidic chips with holes through the plates were placed on a membrane. The medium was flowed through each hole in the membrane as the cells were retained inside each well. Inside these wells, cells formed spheroids and maintained elevated liver function. Microchannels have also been used as improved versions of flat plate bioreactors for hepatocyte culture with the ability to control parameters such as shear stress, contact with parenchymal cells, and oxygenation. Future generations of microfluidic reactors provide powerful means of exposing cells to various physiological stimuli. For example, a Braille system has been developed to create physiological flow conditions such as pulsatile flow. In this scheme, computer-actuated metal pins were used to deform PDMS channels to pump and regulate the flow of the fluid within the microchannels. In addition, complex reactors have been fabricated with the aim of recapitulating the multiple organ interactions of the body. For example, compartmentalized microreactors have been generated to mimic the key functions of the lungs, liver, and fat tissues. These animal-on-a-chip devices could be used to test the pharmacological and toxicity effects of drugs in a controlled manner.

Cell Assembly for Tissue Engineering

Artificial microtissues can also be fabricated by inducing the reaggregation of one or more cell types. These microtissues may be beneficial for applications such as pancreatic, liver, vascular, and cardiac tissue engineering, as well as drug discovery. Layering of cells has been used to engineer myocardial tissues by assembling multiple sheets of cardiomyocytes or to engineer blood vessels by fabricating cylindrically rolled sheets of endothelial cells. Although such approaches may be suitable for some tissue engineering applications, they lack the complexity associated with the architecture of more complex organs. Microscale approaches may provide a solution to this challenge as templates to generate microtissues in a reproducible manner.

For example, by using a combination of microcontact printing and micromachining, hepatocyte spheroids have been formed. More recently, nonadhesive PEG microwells have been used as templates for formation of aggregates of various cell types, including ES cells. This approach aims to overcome the disadvantages associated with the hanging drop and suspension culture methods by providing control over the size, shape, and other features of the cellular assembly in a scalable manner.

The controlled formation of embryoid bodies may also be important in generating more homogeneous cultures that are capable of directing the differentiation of ES cells. In addition, template-based assembly of cells could be used to organize multiple cell types into specific geometries relative to each other within these aggregated tissue sections. It is envisioned that with the integration of such technologies with biomaterials such as photocrosslinkable gels and microfluidics, more complex tissue sections for therapeutic applications can be fabricated.

CONTROLLING THE CELLULAR MICROENVIRONMENT IN VITRO

A major challenge to the clinical feasibility of tissue engineering is a viable cell source. Cells have been traditionally derived from autografts (from the same patient), allografts (from others), or xenografts (from other species). With the emergence of adult stem cells and ES cells, potentially powerful cell sources have emerged. One of the major challenges associated with the use of stem cells is to understand the microenvironmental cues that regulate their fate. Microscale approaches can also be used to control culture conditions and perform high-throughput experimentation, hence providing a suitable tool to study cell–microenvironment interactions *in vitro*. In most tissue culture systems, the cellular microenvironment is vastly different relative to *in vivo* conditions. In the body, cells are exposed to a controlled microenvironment that is tightly regulated with respect to interactions with the surrounding cells, soluble factors, and ECM molecules. In particular, the spatial and temporal distribution of these signals is tightly controlled and unique to each organ. Furthermore, cells in the body are exposed to a 3D environment instead of the 2D environment experienced by cells in traditional culture dishes.

Extracellular cues are important in regulating adult stem cell and ES cell fate decisions. Adult stem cells are present within unique stem cell niches that regulate their self-renewal and differentiation. Engineering artificial stem cell niches, potentially through microscale techniques, is a potent bioengineering approach to regulate stem cell fates *in vitro* . Furthermore, ES cells differentiate based on a series of spatially and temporally regulated signals. Therefore, given the proper series of signals, it may be possible to differentiate ES cells into any desired cell type efficiently and reproducibly. Microscale technologies provide a powerful tool to investigate the extracellular signals that regulate cell fate, because they can control cell–microenvironment interactions and be merged with high-throughput technologies to test many environmental factors simultaneously. In general, microscale approaches to control cell–microenvironment interactions can be separated into cell–cell, cell–matrix, and cell-soluble factor components, which are described below.

Arrays of Cells for Clonal Analysis and Controlling Cell Shape

Cell arrays have been used to pattern stem cells on 2D substrates. Arrays of cells can be used to localize and track individual cells, enabling clonal analysis of stem cell fates. For example, clonal populations of neural stem cells were immobilized within microfabricated structures, and their progeny was tracked by using real-time microscopy, yielding information about cellular kinetics and cell fate decisions in a high-throughput manner. By using this approach, it is possible to study the response of individual stem cells to various microenvironmental signals.

Cell patterning on geometrically defined shapes has been used to study the effects of cell shape on various cell fate decisions. As cells adhere onto micropatterned substrates, they align themselves to the shape of the underlying adhesive region. This shape change induces changes in the cytoskeletal features and has been shown to influence apoptosis and proliferation. More recently, it has been determined that cell shape can also control stem cell differentiation. When human mesenchymal stem cells were patterned on fibronectin islands of various sizes, cells on large islands adhered and flattened, whereas cells on small islands generated spherically shaped cells. As these cells were stained for differentiation markers, it was observed that the spread cells generated osteoblasts, whereas spherical cells gave rise to adipocytes. Further elucidation of the mechanism indicated that cell shape regulated the activation of the RhoA pathway, demonstrating that mechanical stresses experienced during differentiation can be crucial for directing stem cell differentiation. Therefore, controlling the cellular microenvironment through micropatterning may be used for directing cell fate for tissue engineering applications.

Microfluidics for Regulating Cell-Soluble Factor Interactions

The ability to laminarly flow fluids within microchannels can be used to control the spatial positioning of soluble factors relative to cells. Laminar flows form within microchannels because of low convective mixing, which limits molecular transport to diffusion. Laminarly flowing fluids have been used to pattern cells and their microenvironments. Furthermore, an individual cell can be exposed to multiple microenvironments simultaneously by being placed at the interface between two or more adjacent streams. One can treat distinct parts of a cell or cell aggregate to probe the influence of asymmetrically oriented signals on cell behavior or to stain specific regions of a cell for the study of intracellular molecular kinetics.

The partial treatment of cell aggregates has already been used to perturb embryonic development and to study the effects of changing the temperature of part of an embryo. In addition, microfluidic gradient generators that are fabricated by using the same principles have been used to study the effects of concentration gradients on cells. Linear or complex soluble gradients are generated by sequential merging, mixing, and splitting of two or more inlet streams, each of which may contain a particular environmental stimulus. These stable gradients have been used to study complex biological systems. For example, cell chemotaxis in response to soluble chemoattractants has been studied to yield insights into neutrophils migration in response to various shape gradients of IL-8.

Microfluidic gradients have also been used to study neural axon extensions and neural stem cell differentiation. It is important to consider design parameters such as shear stress, transport phenomenon, and material–cell interactions within these devices; however, with the explosion of research in this area in the past few years, such parameters are continuously optimized and modified to attain optimum conditions for various cell types.

Patterned Cocultures for Controlling Cell–Cell Interactions

Inside the body, cells lie in contact or in close proximity to other cell types in a tightly controlled architecture. Tissue engineering constructs that aim to reproduce the architecture and the geometry of tissues will benefit from methods of controlling cell–cell interactions within these tissues. Patterned cocultures are a useful tool for tissue engineered constructs and for studying cell–cell interactions *in vitro,* because they can be used to control the degree of homotypic and heterotypic cell–cell contact. Some of the pioneering work in this area has been performed by studying the interaction of hepatocytes and nonparenchymal fibroblasts in cocultures.

By using these techniques, specific information about the nature of hepatoycte and fibroblast interaction in culture has been elucidated. Although the original methods of fabricating patterned coculture required selective adhesion of each cell types to a particular material on a patterned region, subsequent methods have been developed that allow a wider array of cell types to be tested. The recently developed methods are based on thermally reversible polymers, layer-by-layer deposition of ionic polymers, microfluidic deposition, and molding of hydrogels. Through these approaches, other coculture techniques have been used to probe the interaction of a number of other cell types, including ES cells and other developmentally relevant cell types.

HIGH-THROUGHPUT ASSAYS FOR TISSUE ENGINEERING

Material arrays

As methods to develop small-molecule, polymeric, and genome-based libraries become more widespread and available, there is a great need to test these libraries by using high-throughput assays. Such libraries are useful for tissue engineering and have already yielded candidates that have been shown to induce osteogenesis and cardiomyogenesis from ES cells as well as the dedifferentiation of committed cells. Microscale technologies can miniaturize assays and facilitate high-throughput experimentation and therefore provide a potential tool for screening libraries. Recently, robotic spotters capable of dispensing and immobilizing nanoliters of material have been used to fabricate microarrays, where cell–matrix interactions can be tested and optimized in a high-throughput manner.

For example, synthetic biomaterial arrays have been fabricated to test the interaction of stem cells with various extracellular signals. In this approach, thousands of polymeric materials were synthesized, and their effect on differentiation of human ES cells and human mesenchymal stem cells was evaluated. These interactions have led to unexpected cell–material interactions. Although the molecular mechanisms associated with the biological responses have yet to be clarified, such technology may be widely applicable in cell–microenvironment studies and in the identification of cues that induce desired cell responses. Also, the materials that yield the desired responses could be used as templates for tissue engineering scaffolds. Such an approach is a radical change from traditional methods of developing new biomaterials, where polymers have been individually developed and tested for their effect on cells. In addition to analyzing synthetic material libraries, the effect of natural ECM molecules on cell fate can be evaluated in a high-throughput manner. In one example, combinatorial matrices of various natural ECM proteins were tested for their ability to maintain the function of differentiated hepatocytes and to induce hepatic differentiation from murine ES cells.

Microfluidic systems

Microfluidic systems are ideal for performing cell-based high-throughput experiments because they are cheap and can be used to miniaturize assays, reduce expensive reagent volumes, and reduce cell numbers. Microfluidic arrays can be used to perform screening experiments, to test drug toxicity, and to optimize culture conditions for inducing specific cell fates. To create such systems, it is important to be able to culture cells inside microfluidic devices in a reliable and long-term manner and to be able to combine these devices in such a way as to test many conditions simultaneously. Many experiments have localized cells within microfluidic channels; however, the long-term survival of cells inside these channels has not been demonstrated. In addition, it is important to localize the cells in particular regions of microfluidic channels by using substrate patterning or topographical structures.

Also, it is important to interface the microscale channels to the outside world. Current techniques include the use of complicated valves and pumps to fabricate large arrays of microchannels that can be individually addressed. The ability to test various combinations and doses of drugs can be useful for performing factorial experiments, which can optimize culture conditions for inducing various biological fates. An alternative approach for cell-based molecular screening is to use multiphenotype cell arrays. In these systems, the effect of a soluble factor can be tested on many cell types simultaneously. Incorporating multiphenotype cell arrays within microfluidic arrays has shown promise as a tool for fabricating high-throughput assays.

REGULATION OF 3D STEM CELL MICROENVIRONMENT IN TISSUE REGENERATION

CALL FOR PAPERS

Stem cells, including both pluripotent and adult stem cells, have significant potential for use in tissue regeneration. Regardless of their origins, common properties shared by stem cells are their ability to undergo self-renewal and lineage-specification under certain physiologic or experimental conditions.

During development, embryonic stem cells give rise to all the specialized cells within the body, a property that has also been demonstrated by induced pluripotent stem cells. In adult organisms, stem or progenitor cells reside in tissue-specific microenvironments and act as the repair system that maintains the integrity of the tissue through the lifespan of the organism. These properties make stem cells attractive in tissue regeneration to replace human cells, tissues, or organs to restore or maintain normal tissue function.

It is well established that the stem cell microenvironments significantly influence cell fate decision and play essential roles in their development into organized tissue structures. Key components of the microenvironments in which stem cells develop or reside include soluble factors, extracellular matrix molecules, neighboring cells, and physical and physicochemical stimuli. Recent advanced studies of both pluripotent and adult stem cells utilizing novel technologies have identified these components and revealed their regulatory roles in directing stem cell fate.

Tissue regeneration using stem cells requires the formation of 3D tissue architectures. In this process, recapitulating the microenvironmental components of the stem cells and their interactions has become an important approach in the development of 3D constructs. Recent advances in biomaterials, scaffold-free construct fabrication technology, and bioreactors have provided powerful tools to regulate stem cell fate and direct their developmental process.

The main focus of this special issue will be on the technological innovations and development related to the regulation of stem cell microenvironment in 3D structures in tissue regeneration. We invite researchers to contribute original research as well as review articles that improve our understanding of stem cell 3D microenvironment and the application of the knowledge in 3D constructs. Potential topics include, but are not limited to:

- Novel biomaterials for controlling stem cell fate and 3D organization
- Scaffold-free approaches to engineering 3D tissues
- Regulation of stem cell microenvironment in bioreactors
- Spatiotemporal cues for developing composite tissues
- Influence of mechanics on stem cell fate

Future Directions

The synthesis of new materials and the incorporation of improved fabrication strategies have led to dramatic enhancements in the complexity and biomimecry of today's tissue engineering constructs. In particular, the use of microscale technologies in tissue engineering, either directly as transplantable constructs or indirectly as a tool for understanding the biology of organs, has pushed the field closer to clinical therapies. However, significant challenges need to be addressed. These barriers include the lack of suitable materials with the desired degradation rates, products, and suitable mechanical properties for the desired tissue. Another challenge that must be addressed is the optimization of scaffold architecture, including pore size, morphology, surface topography, and bioactivity. Also, new and optimized processing methods must be developed to address issues related to cell seeding, vascularization, and fabrication of 3D tissues without cumbersome methods of assembly. Furthermore, more research is required in testing and validating the *in vivo* functionality of microfabricated constructs and in assessing the performance of these constructs against competing technologies.

In addition, microscale tools have provided much insight into the fundamental biology of how cells interact with the surrounding components, such as cell–cell, cell-soluble factors, and cell–ECM molecules. This knowledge should be used to direct cell fates as a cell source or be incorporated into tissue engineering scaffolds. Thus, the continued merger of engineering, medicine, materials, and biological sciences as mediated by microscale approaches in tissue engineering and biology will enhance our ability to create *in vivo*-like physiological models that can be used for fabricating tissues or for understanding fundamental biology.

Index